SEMINARS IN MATHEMATICS
V. A. Steklov Mathematical Institute, Leningrad

Volume 17

MATHEMATICAL PROBLEMS IN WAVE PROPAGATION THEORY

Part III

Edited by V. M. Babich

Translated from Russian
by James S. Wood

CONSULTANTS BUREAU • NEW YORK – LONDON • 1972

The original Russian text was published by Nauka Press in Leningrad in 1970 by offset reproduction of manuscript. The hand-written symbols have been retained in this edition. The present translation is published under an agreement with Mezhdunarodnaya Kniga, the Soviet book export agency.

В.М.БАБИЧ

МАТЕМАТИЧЕСКИЕ ВОПРОСЫ ТЕОРИИ РАСПРОСТРАНЕНИЯ ВОЛН.3.

MATEMATICHESKIE VOPROSY TEORII RASPROSTRANENIYA VOLN. 3

Library of Congress Catalog Card Number 79-13851
ISBN 0-306-18817-1

A Division of Plenum Publishing Corporation
227 West 17th Street, New York, N. Y. 10011

United Kingdom edition published by Consultants Bureau, London
A Division of Plenum Publishing Company, Ltd.
Davis House (4th Floor), 8 Scrubs Lane, Harlesden, London, NW10 6SE, England

Printed in the United States of America

SEMINARS IN MATHEMATICS
V. A. STEKLOV MATHEMATICAL INSTITUTE, LENINGRAD

ZAPISKI NAUCHNYKH SEMINAROV
LENINGRADSKOGO OTDELENIYA
MATEMATICHESKOGO INSTITUTA IM. V. A. STEKLOVA AN SSSR

ЗАПИСКИ НАУЧНЫХ СЕМИНАРОВ
ЛЕНИНГРАДСКОГО ОТДЕЛЕНИЯ
МАТЕМАТИЧЕСКОГО ИНСТИТУТА им. В.А. СТЕКЛОВА АН СССР

SEMINARS IN MATHEMATICS
V. A. Steklov Mathematical Institute, Leningrad

1	Studies in Number Theory	A. V. Malyshev, Editor
2	Convex Polyhedra with Regular Faces	V. A. Zalgaller
3	Potential Theory and Function Theory for Irregular Regions	Yu. D. Burago and V. G. Maz'ya
4	Studies in Constructive Mathematics and Mathematical Logic, Part I	A. O. Slisenko, Editor
5	Boundary Value Problems of Mathematical Physics and Related Aspects of Function Theory, Part I	V. P. Il'in, Editor
6	Kinematic Spaces	R. I. Pimenov
7	Boundary Value Problems of Mathematical Physics and Related Aspects of Function Theory, Part II	O. A. Ladyzhenskaya, Editor
8	Studies in Constructive Mathematics and Mathematical Logic, Part II	A. O. Slisenko, Editor
9	Mathematical Problems in Wave Propagation Theory, Part I	V. M. Babich, Editor
10	Isoperimetric Inequalities in the Theory of Surfaces of Bounded External Curvàture	Yu. D. Burago
11	Boundary Value Problems of Mathematical Physics and Related Aspects of Function Theory, Part III	O. A. Ladyzhenskaya, Editor
12	Investigations in the Theory of Stochastic Processes	V. N. Sudakov, Editor
13	Investigations in Classical Problems of Probability Theory and Mathematical Statistics, Part I	V. M. Kalinin and O. V. Shalaevskii
14	Boundary Value Problems of Mathematical Physics and Related Aspects of Function Theory, Part IV	O. A. Ladyzhenskaya, Editor
15	Mathematical Problems in Wave Propagation Theory, Part II	V. M. Babich, Editor
16	Studies in Constructive Mathematics and Mathematical Logic, Part III	A. O. Slisenko, Editor
17	Mathematical Problems in Wave Propagation Theory, Part III	V. M. Babich, Editor
18	Automatic Programming and Numerical Methods of Analysis	V. N. Faddeeva, Editor
19	Investigations in Linear Operators and Function Theory, Part I	N. K. Nikol'skii, Editor
20	Investigations in Constructive Mathematics and Mathematical Logic, Part IV	Yu. V. Matiyasevich and A. O. Slisenko, Editors
21	Boundary Value Problems of Mathematical Physics and Related Aspects of Function Theory, Part V	O. A. Ladyzhenskaya, Editor
22	Investigations in Linear Operators and Function Theory, Part II	N. K. Nikol'skii, Editor
23	Numerical Methods and Functional Analysis	S. G. Mikhlin and V. P. Il'in, Editors
24	External Problems in the Geometric Theory of Functions with Complex Variables	Yu. G. Alenitsyn and G. B. Kuz'minii, Editors

PREFACE

The subject matter of the articles in the present collection reflects the thematic content of the LOMI AN SSR (Leningrad Section of the V. A. Steklov Mathematical Institute, Academy of Sciences of the USSR) Seminar on the Mathematical Problems of Wave Diffraction and Propagation Theory. The bulk of the Seminar's research is aimed at asymptotic methods, to which the articles presented in this book are oriented in one form or another, with the exception of the articles by L. A. Antonova, R. G. Barantsev, and V. V. Kozachek (numerical methods) and by B. P. Belinskii, D. P. Kouzov, V. D. Luk'yanov, and V. D. Chel'tsova (formulation of the solution of a particular problem in explicit form).

V. M. Babich

CONTENTS

NUMERICAL SOLUTION OF THE SCATTERING PROBLEM FOR BODIES OF REVOLUTION

L. A. Antonova, R. G. Barantsev, and V. V. Kozachek

We wish to consider axisymmetric scattering by a finite body bounded by a Lyapunov surface σ. The wave function ψ, satisfying the equation

$$\Delta \Psi + \kappa^2 \Psi = 0 \tag{1}$$

for $r \geq h = \max_{\sigma} \{ r \}$, is expanded into a series:

$$\Psi(r, \vartheta) = \sum_{n=0}^{\infty} \left[-c_n^- X_n^-(r) + c_n^+ X_n^+(r) \right] Y_n(\vartheta), \tag{2}$$

where

$$Y_n(\vartheta) = \sqrt{\frac{2n+1}{2}} P_n(\cos\vartheta), \quad X_n^{\pm}(r) = \sqrt{\frac{\pi}{2\kappa r}} H_{n+\frac{1}{2}}^{(1)(2)}(\kappa r),$$

P_n are Legendre polynomials, and $H_{n+\frac{1}{2}}$ are Hankel functions.

We denote

$$\varphi_n^{\pm} = X_n^{\pm} Y_n, \quad \varphi_n^{0} = \frac{\varphi_n^- + \varphi_n^+}{2}. \tag{3}$$

For large r the waves φ_n^- yield a negative radial flux, and φ_n^+ is positive. The scattering problem entails the determination of the amplitudes c_n^+ of the outgoing waves when the amplitudes c_n^- of the incoming waves and a certain condition on the surface σ are specified.

Let

$$\Psi|_{\sigma} = 0. \tag{4}$$

Then (see [1])

$$c_n^{\pm} = (q, \varphi_n^{\pm}), \quad n = 0, 1, 2, \ldots, \tag{5}$$

*An expanded portion of a paper presented at the Sixth All-Union Acoustic Conference in Moscow, 1968.

where

$$q = \frac{\kappa}{2i} \frac{\partial \psi}{\partial n}\Big|_{\sigma},$$

$$(q, \varphi) = \iint_{\sigma} q \varphi^{*} d\sigma, \qquad \varphi^{\pm} = \varphi^{\mp *}.$$

When only one unit-amplitude wave, say the mth, arrives, i.e., when

$$c_n^- = \delta_{mn}, \qquad n = 0, 1, 2, \ldots, \tag{6}$$

the solution for $r \geq h$ has the form

$$\psi_m = \sum_{n=0}^{\infty} (-\delta_{mn} \varphi_n^- + s_{mn} \varphi_n^+), \tag{7}$$

where

$$s_{mn} = (q_m, \varphi_n^+), \tag{8}$$

$$\delta_{mn} = (q_m, \varphi_n^-). \tag{9}$$

Forming

$$\psi = \sum_{m=0}^{\infty} c_m^- \psi_m, \tag{10}$$

and changing the order of summation, we arrive at (2), so that

$$c_n^+ = \sum_{m=0}^{\infty} c_m^- S_{mn}, \qquad n = 0, 1, 2, \ldots. \tag{11}$$

The matrix $S = \{S_{mn}\}$ transforms the incoming-wave amplitudes c_m^- into the outgoing-wave amplitudes c_n^+, i.e., solves the scattering problem. The element s_{mn} is interpreted as the coefficient of transformation of the m th incoming wave into the n th outgoing wave. When κ is fixed, the scattering matrix is completely determined by the shape of the body and the boundary condition on it. Knowing S, we can obtain the scattered field for any incident wave.

As Eq. (5) indicates, in order to find S it is required to expand $\varphi_n^-|_{\sigma}$ in terms of $\varphi_m^+|_{\sigma}$. It is well known [2, 3] that the system of functions φ_n^+ is complete in $\mathcal{L}_2$ on a Lyapunov surface.

In this case

$$\iint_{\sigma} \Big| \varphi_n^- - \sum_{m=0}^{N} \varphi_m^+ s_{mn}^{(N)} \Big|^2 d\sigma \xrightarrow[N \to \infty]{} 0, \tag{12}$$

where $s_{mn}^{(N)}$ are determined from the equations

$$\sum_{m=0}^{N} (\varphi_l^+, \varphi_m^+) s_{mn}^{(N)} = (\varphi_l^-, \varphi_n^+), \quad l = 0, 1, \ldots, N. \tag{13}$$

It is readily shown that

$$s_{mn}^{(N)} \xrightarrow[N \to \infty]{} s_{mn}.$$

The matrix S is unitary and symmetric [4]. We have carried out concrete numerical calculations for ellipsoids of revolution characterized by semiaxes a and b and for nonconvex bodies formed by the revolution of a lemniscate:

$$r(\vartheta) = \sqrt{1+\varepsilon^2 \cos^2\vartheta}, \tag{14}$$

where

$$\varepsilon^2 = 1 - \frac{a^2}{b^2} \tag{15}$$

We obtain the following values for the parameters:

$$a/b = 1{,}5, \qquad ka = 1{,}31;\ 2;\ 3;\ 5.$$

We also calculated the following case for an ellipsoid of revolution:

$$a/b = 2, \qquad ka = 1.31.$$

We calculated the scattering matrices, plane-wave scattering amplitudes, and the differential and total scattering cross sections. The results of the calculations are summarized in Tables 1, 2, and 3.

We then plotted graphs (see Figs. 1-3) of the normalized scattering function

$$T_w = \frac{\mathcal{I}(\vartheta)}{\sigma}, \tag{16}$$

where

$$\mathcal{I}(\vartheta) = |f(\vartheta)|^2, \tag{17}$$

$$\mathcal{J}_{em} = \iint_\sigma \varphi_e^- \varphi_m^+ \, ds. \tag{18}$$

We carried out the calculations for two bodies, an ellipsoid of revolution and a body formed by revolution of a lemniscate, in order that we might explicate the limits of practical applicability of the method. With this object in mind we chose a broad interval of variation of the numbers k. The principal computational difficulties were encountered in the evaluation of integrals of the type

$$f(\vartheta) = \frac{1}{k} \sum_{n=0}^{\infty} (c_n^- + c_n^+)\, i^{n-1} \sqrt{\frac{2n+1}{2}}\, P_n(\cos\vartheta). \tag{19}$$

The integrand oscillates very rapidly for large values of the indices l and m. So as not to diminish accuracy we calculated the $\mathcal{J}_{l+1/2}$ (see [5]) according to retrograde recursion formulas (from large to small indices). We computed integrals of the type (15) in the system (13) according to area-of-rectangle formulas with a relative error $\varepsilon = 10^{5}$. The solution of the system (13) did not present any difficulties. The accuracy of approximation of the matrix $S^{(N)}$ to S was determined from the realization of symmetry and unitarity of the matrix $S^{(N)}$. These properties were not utilized in the calculations and thus afforded a good criterion of the reliability of the end results. The calculations were run on M-20 and BESM-3M digital computers. The machine time for calculation of one variant ranged from 6 to 50 min, the latter for $ka = 5$. The calculations indicated applicability of the method for elongations a/b of the order of 1 or 2 and ka on the order of 1 to 5. The restriction imposed by the growth of the parameter ka is determined by the growth of the required machine time, and it is possible in general, as demonstrated by preliminary calculations, to attain values of ka of the order 10. We made some attempts to work toward larger elongations a/b. When a/b was made greater than 2 it was impossible to arrive at a symmetric scattering matrix, even with $ka = 1.31$ and sufficiently large values of N. It seems likely that the method fails at large elongations, in which case some additional study is indicated.

TABLE 1. Scattering Matrices for Scattering by an Ellipsoid

a/b = I.5 , κa = I.3I

$$\begin{pmatrix} 0.900-0.428\,i & 0 & 0.065+0.038\,i & 0 & 0.00I \\ 0 & -0.847+0.53I\,i & 0 & 0.004+0.0I3\,i & 0 \\ 0.065+0.038\,i & 0 & -0.996+0.055\,i & 0 & 0.00I \\ 0 & 0.004+0.0I3\,i & 0 & -I+0.002\,i & 0 \\ 0.00I & 0 & 0 & 0 & -I \end{pmatrix}$$

a/b = I.5 κa = 2

$$\begin{pmatrix} 0.976+0.086\,i & 0 & 0.I96-0.02I\,i & 0 & 0.005-0.00I\,i \\ 0 & -0.237+0.969\,i & 0 & 0.044+0.054\,i & 0 \\ 0.I97+0.022\,i & 0 & -0.937+0.28I\,i & 0 & 0.002+0.0I0\,i \\ 0 & 0.045+0.055\,i & 0 & -0.997+0.032\,i & 0 \\ 0.005+0.00I\,i & 0 & 0.002+0.0II\,i & 0 & -I+0.00I\,i \end{pmatrix}$$

a/b = I.5 , κa = 3

$$\begin{pmatrix} 0.I6I-0.904\,i & 0 & 0.I25-0.373\,i & 0 & 0.004-0.027\,i & 0 & -0.00I \\ 0 & -0.878-0.429\,i & 0 & 0.247+0.2I3\,i & 0 & 0.0II & 0 \\ 0.I25+0.373\,i & 0 & -0.4I9+0.8I3\,i & 0 & 0.05I+0.07I\,i & 0 & 0.00I+0.002\,i \\ 0 & 0.247+0.02I\,i & 0 & -0.929+0.272\,i & 0 & 0.004+0.0I8\,i & 0 \\ 0.004-0.028\,i & 0 & 0.052+0.072\,i & 0 & -I+0.040\,i & 0 & 0.002 \\ 0 & 0.0II & 0 & 0.004+0.0I8\,i & 0 & -I.+0.003\,i & 0 \\ 0.00I & 0 & 0.002+0.002\,i & 0 & 0.002 & 0 & -I \end{pmatrix}$$

a/b = 2 , κa = I.3I

$$\begin{pmatrix} 0.II+0.99\,i & 0 & 0.06+0.05\,i & 0 \\ 0 & 0.9I+0.4I\,i & 0 & 0.0I \\ 0.06+0.05\,i & 0 & -I+0.03\,i & 0 \\ 0 & 0.003+0.00I\,i & 0 & -I \end{pmatrix}$$

a/b = I.5 , κa = 5

$$\begin{pmatrix} -0.29+0.66\,i & 0 & -0.46+0.48\,i & 0 & -0.I4+0.I3\,i & 0 & -0.0I+0.0I\,i & 0 \\ 0 & -0.63-0.44\,i & 0 & -0.37-0.5I\,i & 0 & -0.08+0.08\,i & 0 & -0.0I-0.0I\,i \\ -0.46+0.48\,i & 0 & 0.55-0.24\,i & 0 & 0.35-0.26\,i & 0 & 0.03-0.03\,i & 0 \\ 0 & -0.38-0.5I\,i & 0 & 0.I9+0.70\,i & 0 & 0.23+0.II\,i & 0 & 0.0I+0.0I\,i \\ -0.I4+0.I3\,i & 0 & 0.35-0.27\,i & 0 & -0.62+0.6I\,i & 0 & 0.05+0.09\,i & 0 \\ 0 & -0.08-0.08\,i & 0 & 0.24+0.0II\,i & 0 & -0.93+0.22\,i & 0 & 0.0I+0.03\,i \\ -0.0I+0.0I\,i & 0 & 0.03-0.04\,i & 0 & 0.05+0.09\,i & 0 & -0.99+0.04\,i & 0 \\ 0 & -0.0I-0.0I\,i & 0 & 0.02+0.0I\,i & 0 & 0.0I+0.03\,i & 0 & -I \end{pmatrix}$$

TABLE 2. Scattering Matrices for Scattering by the Body of Revolution $r(\vartheta) = \sqrt{1+\varepsilon^2\cos^2\vartheta}$

a/b = I.5

ka = I.3I

$$\begin{pmatrix}
0.5I5+0.853\,i & 0 & 0.075+0.039\,i & 0 & 0 \\
0 & -0.8I5+0.579\,i & 0 & 0.003+0.0II\,i & 0 \\
0.075+0.039\,i & 0 & -0.994+0.06I\,i & 0 & 0 \\
0 & 0.004+0.0I2\,i & 0 & -I & 0 \\
0 & 0 & 0 & 0 & -I
\end{pmatrix}$$

a/b = I.5

ka = 2

$$\begin{pmatrix}
0.974-0.059\,i & & 0.2I4-0.043\,i & 0 & 0.00I-0.00I\,i & 0 \\
0 & -0,II3+0.99I\,i & 0 & 0.046+0.050\,i & 0 & 0 \\
0.2I5-0.043\,i & 0 & -0.922+0.3I7\,i & 0 & 0.002+0.009\,i & 0 \\
0 & 0.047+0.050\,i & 0 & -0.997+0.037\,i & 0 & 0 \\
0.00I-0.00I\,i & 0 & 0.002+0.00I\,i & 0 & -I & 0 \\
0 & 0 & 0 & 0 & 0 & -I
\end{pmatrix}$$

$a/b \pm 5$ $\quad ka$ = I.5

$$\begin{pmatrix}
-0.042+0.639i & 0 & -0.359+0.648i & 0 & -0.022+0.20Ii & 0 & 0.004+0.005i & 0 & 0 \\
0 & -0.792-0.II7i & 0 & -0.478-0.350i & C & -0.090+0.00Ii & 0 & -0.003+0.00Ii & 0 \\
-0.359+0.648i & 0 & 0.4I5-0.337i & 0 & 0.300-0.272i & 0 & -0.00I-0.030i & 0 & -0.00I \\
0 & -0.478-0.350i & 0 & 0.33I+0.684i & 0 & 0.245+0.I03i & 0 & 0.0I0+0.00Ii & 0 \\
-0.022+0.20Ii & 0 & 0.300-0.272i & 0 & -0.530+0.708i & 0 & 0.059+0.I03i & 0 & 0.002+0.004 \\
0 & -0.090+0.00Ii & 0 & 0.245+0.I02i & 0 & -0.9I4+0.290i & 0 & 0.006+0.030i & 0 \\
0.004+0.005i & 0 & -0.00I-0.030i & 0 & 0.059+0.I04i & 0 & -0.900-0.067i & 0 & 0.005 \\
0 & -0.003+0.00Ii & 0 & 0.0I0+0.00Ii & 0 & 0.007+0.030i & 0 & -I+0.007i & 0 \\
0 & 0 & -0.00I & 0 & 0.002+0.004i & 0 & 0.005 & 0 & -I.-0.00I
\end{pmatrix}$$

a/b = I.5 , ka = 3

$$\begin{pmatrix}
-0.039-0.900\,i & 0 & 0.074+0.427\,i & 0 & -0.0I6-0.0I2\,i & 0 & 0 \\
0 & 0.947+0.222\,i & 0 & 0.232-0.0I2\,i & 0 & 0.004-0.002\,i & 0 \\
0.074-0.427\,i & 0 & -0.336+0.83I\,i & 0 & 0.05I+0.072\,i & 0 & 0.00I+0.00I\,i \\
0 & 0.232+0.0I2\,i & 0 & -0.920+0.3I5\,i & 0 & 0.004+0.0I9\,i & 0 \\
-0.0I7-0.0I2\,i & 0 & 0.052+0.073\,i & 0 & -0.995+0.048\,i & 0 & 0 \\
0 & 0.004+0.002\,i & 0 & 0.004+0.020\,i & 0 & -I+0.004\,i & 0 \\
0 & 0 & 0.00I+0.00I\,i & 0 & 0.002 & 0 & -I
\end{pmatrix}$$

TABLE 3. Outgoing-Wave Amplitudes for Scattering by Bodies of Revolution

I Ellipsoid a/b = I.5

	ka = I.3I	ka = 2	ka = 3	ka = 5
C_0^+	-0.202-0.576 i	-0.39I-0.092 i	0.075+0.I04 i	-0.260+0.056 i
C_1^+	0.626+I.046 i	I.087+0.372 i	0.486-0.627 i	0.239+0.248 i
C_2^+	-I.62I+0.058 i	-I.625+0.407 i	-0.857+I.404 i	0.527-0.239 i
C_3^+	0.0I2-I.876 i	0.009-I.920 i	-0.44I-2.050 i	-I.67I+0.3I4 i
C_4^+	2.I2I	2.I2I+0.0I3 i	2.20I-0.054 i	2.076-I.608 i
C_5^+		2.344	-0.027+2.339 i	0.I63+2.742 i
C_6^+			-I.546	-2.586-0.I58 i
C_7^+				0.024-2.708 i
C_8^+				2.902-0.007 i
C_9^+				-0.002+3.080 i
$2\pi\sigma$	I3.08	II.43	9.76	9.62

II Body of revolution $r(\vartheta)=\sqrt{1+\varepsilon^2\cos^2\vartheta}$, a/b = I.5

	ka = I.3I	ka = 2	ka = 3	ka = 5
C_0^+	-0.246-0.54I i	-0.352-0.003 i	0.I23-0.0I3 i	-0.48I+0.I59 i
C_1^+	0.688+I.004 i	I.I20+0.224 i	0.290-0.735 i	0.798+0.279 i
C_2^+	-I.625+0.007 i	-I.6I4+0.5I2 i	-0.689+I.466 i	0.27I-0.488 i
C_3^+	0.0I5-I.874 i	-0.008-I.923 i	-0.530-2.0I5 i	-I.469+0.658 i
C_4^+	2.I2I	2.I23+0.0I4 i	2.205+0.022 i	I.759-I.823 i
C_5^+		2.345 i	-0.026+2.348 i	0.408+2.728 i
C_6^+			-2.548-0.003 i	-2.654-0.II5 i
C_7^+			-2.739 i	0.053-2.733 i
C_8^+				2.9II-0.006 i
$2\pi\sigma$	I4.II	I2.89	II.54	I0.83

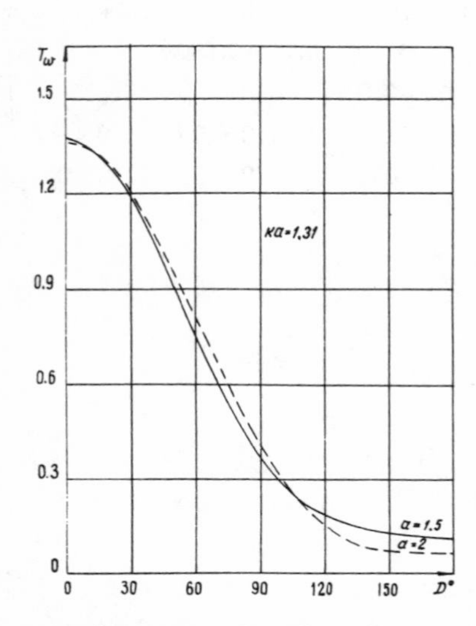

Fig. 1. Normalized scattering function for an ellipsoid.

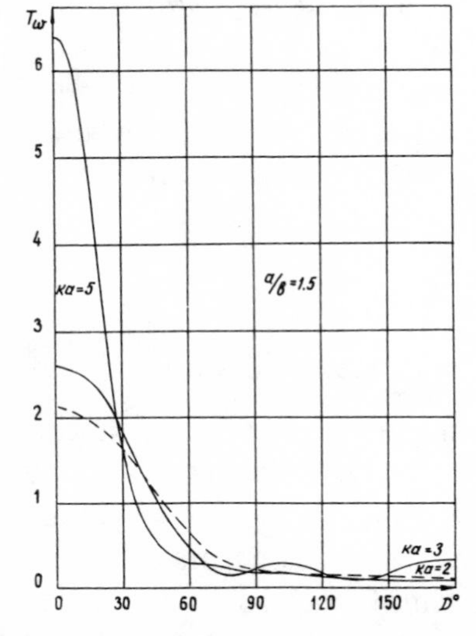

Fig. 2. Normalized scattering function for an ellipsoid.

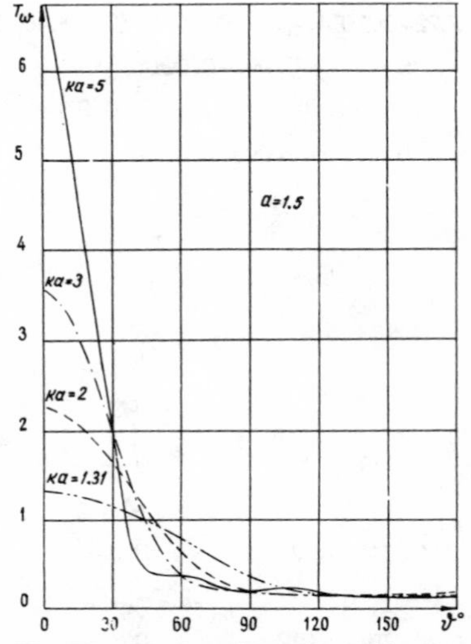

Fig. 3. Normalized scattering function for the body of revolution $r(\vartheta)=\sqrt{1+\varepsilon^2\cos^2\vartheta}$.

The results for one of the bodies formed by revolution of the lemeniscate $r(\vartheta)=\sqrt{1+\varepsilon^2\cos^2\vartheta}$ were compared with the results obtained earlier in [6]. The agreement was good.

The convexity property of the bodies proved nonessential to the calculations.

LITERATURE CITED

1. Barantsev, R. G., Method of separation of variables in the problem of scattering by a body of arbitrary configuration, Dokl. Akad. Nauk SSSR, 147(3):569-570 (1962).
2. Vekua, I. N., On the completeness of a system of metaharmonic functions, Dokl. Akad. Nauk SSSR, 90(5):715-718 (1953).
3. Kravtsov, V. V., Integral equations in diffraction problems, in: Computational Methods and Programming, No. 5, VTs MGU, Moscow (1966), pp. 260-293.
4. Barantsev, R. G., and Kozachek, V. V., Investigation of the matrix of scattering by objects of complex configurations, Vestnik Leningrad Univ. (LGU), 7(2):71-77 (1968).
5. Lozinskii, N. N., and Erglis, V. R., Programmer's Handbook, Vol. 1, Leningrad (1963), p. 208.

FINDING THE SADDLE POINT IN THE ELLIPSE PROBLEM

V. M. Babich

Let a point source of oscillations act at the point ξ_0, η_0* on a perfectly reflecting ellipse $\xi = \xi_0$, $0 \le \eta < 2\pi$. The problem of determining the high-frequency asymptotic behavior of the wave field at a certain point (Ξ, H) in the domain subtended by irradiation from (ξ_0, η_0) is solved by applying the method of steepest descents to an integral of the form

$$\int_{\mathcal{L}} F(\tau) e^{i\kappa\omega(\tau)} d\tau, \quad \kappa \gg 1. \tag{1}$$

Here κ is the wave number, $F(\tau)$ is a regular function of τ independent of κ, and $\mathcal{L}$ is a contour on the plane of the complex variable τ.

The phase function in the integral (1) has the form

$$\omega(\tau) = c\int_{\xi_0}^{\Xi} (\mathrm{ch}^2\xi - \tau^2)^{\frac{1}{2}} d\xi + c\int_{\eta_0}^{H} (\tau^2 - \cos^2\eta)^{\frac{1}{2}} d\eta \tag{2}$$

(see [1, 2, 3]).

In order to find the saddle point in the integral (1) it is necessary to solve the equation $\omega'(\tau) = 0$. It was first solved by Andronov [3], who indicated the following very elementary geometric technique for constructing the real root of this equation: we draw a line ℓ connecting the points $M(\Xi\ H)$ and $M_0(\xi_0, \eta_0)$. We do not let ℓ cross the segment between the foci of the ellipse $\xi = \xi_0$ (see Fig. 1). We construct the ellipse confocal with the ellipse $\xi = \xi_0$ and tangent to ℓ. The length of the semimajor axis of this ellipse is then the unique real root of the function $\omega'(\tau) = 0$. A similar rule can be formulated in the case when the line ℓ crosses the segment between the foci of the ellipse $\xi = \xi_0$. In this case the role of the confocal ellipse is taken by the hyperbola confocal with $\xi = \xi$ and tangent to ℓ. The length of the real semiaxis of this hyperbola is then equal to the desired root of the equation $\omega'(\tau) = 0$.

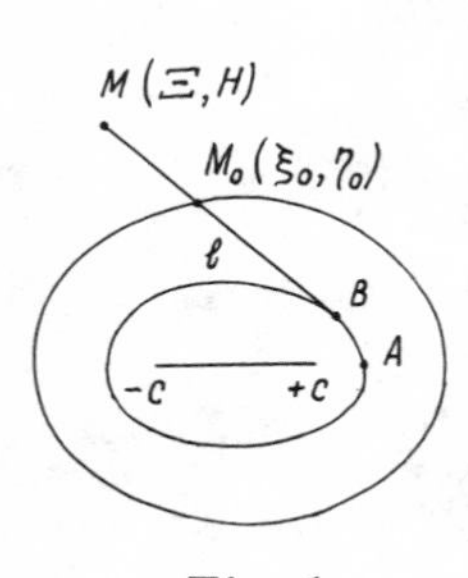

Fig. 1.

*The ensuing analysis is carried out exclusively in the elliptic coordinate system (ξ, η), which is related to the Cartesian system (x_1, x_2) by the usual equations

$$x_1 = c\,\mathrm{ch}\,\xi \cos\eta, \quad x_2 = c\,\mathrm{ch}\,\xi \sin\eta, \quad c > 0.$$

The proof given by Andronov for the foregoing propositions is based on the addition theorems for elliptic integrals. It is exceedingly intricate, and its generalization to other diffraction problems in which the solution can be constructed by the method of separation of variables is very difficult. Nevertheless, appropriate calculations have shown that the saddle points in problems related to diffraction by a circular disk, spherical ball [4], parabola, or paraboloid of revolution are determined by a technique entirely analogous to the one just described.

Our objective is to give a proof of the above geometric construction of the root of the equation $\omega'(\tau)=0$, such that excessive computations are not required and yet the proof can be extended to analogous diffraction problems.

We confine our present analysis to the case in which the line $\mathcal{C}$ does not cross the segment between foci of the ellipse $\xi=\xi_0$ (see Fig. 1). We set $\tau=\operatorname{ch}\xi_1$ and represent the phase function $\omega(\tau)$ in the form

$$\omega=\omega_1-\omega_2, \tag{3}$$

where

$$\begin{aligned} \omega_1(\Xi,H)&=c\int_{\xi_1}^{\Xi}(\operatorname{ch}^2\xi-\operatorname{ch}^2\xi_1)^{\frac{1}{2}}d\xi+c\int_0^H(\operatorname{ch}\xi_1-\cos^2\eta)^{\frac{1}{2}}d\eta,\\ \omega_2(\xi_0,\eta_0)&=c\int_{\xi_1}^{\xi_0}(\operatorname{ch}^2\xi-\operatorname{ch}^2\xi_1)^{\frac{1}{2}}d\xi+c\int_0^{\eta_0}(\operatorname{ch}^2\xi_1-\cos^2\eta)^{\frac{1}{2}}d\eta \end{aligned} \tag{4}$$

[see Eq. (2)].

It is readily verified that $\omega_1(\Xi,H)$ [or $\omega_2(\xi_0,\eta_0)$] as a function of the point Ξ, H (or the point ξ_0,η_0) is a solution of the eikonal equation (see, e.g., [6]) in the elliptic coordinate system ξ, η:

$$\left(\frac{\partial\Omega}{\partial\xi}\right)^2+\left(\frac{\partial\Omega}{\partial\eta}\right)^2=c^2(\operatorname{ch}^2\xi-\cos^2\eta)\ . \tag{5}$$

The ray field corresponding to the eikonals ω_1 and ω_2 (see [5]), we readily perceive, represents a set of semitangents to the ellipse $\xi=\xi_1$ (see Fig. 2; the arrows indicate the directions in which ω_i increase).

Thus, for $\xi=\xi_1$ the eikonals $\omega_i(B)$ simply reduce (see Fig. 2) to the arclength of the ellipse $\xi=\xi_1$ from the point A $(\xi=\xi_1,\ \eta=0)$, to the point B, so that at points of the ellipse $\xi=\xi_1$ the unit tangent vector is equal to the gradient of ω_i. This implies that the rays corresponding to the eikonal ω_i ($i=1,2$) are tangent to the ellipse $\xi=\xi_1$. The fact that the rays corresponding to ω_i are in fact the semitangents depicted in Fig. 2 is easily deduced from a consideration of the direction of $\operatorname{grad}\omega_i$

Inasmuch as the eikonal equation (5) corresponds to unit rate of change, the value of ω_i at a point M on any ray m (see Fig. 2) is equal to the length of the segment BM, up to a term that is constant on this ray. The indicated term is equal at B to the arclength of AB [see Fig. 2 and Eq. (4)], so that the equation for the eikonal ω_i is equal at M to the length of the line ABM (see Fig. 2).

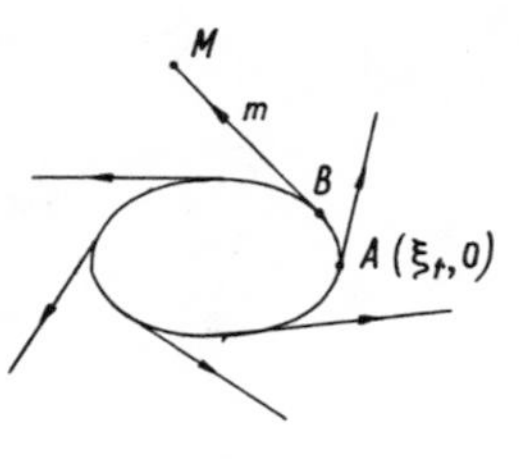

Fig. 2

Next we verify the fact that τ, constructed by Andronov's method is indeed the root of the equation

$$\omega'(\tau)=0.$$

The latter is equivalent to the following proposition: if the ellipse $\xi=\xi_1$ is constructed as indicated in Fig. 1, then for this particular ξ_1

$$\frac{\partial \omega}{\partial \xi_1} = \frac{\partial}{\partial \xi_1}\left[c\int_{\xi_0}^{\Xi}(\mathrm{ch}^2\xi - \mathrm{ch}^2\xi_1)^{\frac{1}{2}}d\xi + c\int_{\eta_0}^{H}(\mathrm{ch}^2\xi_1 - \cos^2\eta)^{\frac{1}{2}}d\eta\right] = \frac{\partial}{\partial \xi_1}(\omega_1 - \omega_2) = 0. \tag{6}$$

Using the classical equation for the variation of an integral in parametric form with moving endpoints,

$$\delta\int_{M_0}^{M}\mathcal{F}(\dot{x}_1, \dot{x}_2, x_1, x_2)d\sigma = \int_{M_0}^{M}\sum_{i=1}^{2}\left(\mathcal{F}_{x_i} - \frac{d}{d\sigma}\mathcal{F}_{\dot{x}_i}\right)\delta x_i\, d\sigma + \sum_{i=1}^{2}\mathcal{F}_{\dot{x}_i}\,\delta x_i\Big|_{M_0}^{M},$$

$$\mathcal{F}(k\dot{x}_1, k\dot{x}_2, x_1, x_2) \equiv k\mathcal{F}(\dot{x}_1, \dot{x}_2, x_1, x_2), \quad k>0; \quad \dot{x}_i = \frac{dx_i}{d\sigma},$$

we at once infer that

$$\frac{\partial \omega_1}{\partial \xi_1} = \delta\int_{ABM}\sqrt{\dot{x}_1^2 + \dot{x}_2^2}\,d\sigma = \int_{AB}\sum_{i=1}^{2}\left(-\frac{d}{d\sigma}\frac{\partial}{\partial \dot{x}_i}\sqrt{\dot{x}_1^2 + \dot{x}_2^2}\right)dx_i\, d\sigma = (c^2\mathrm{ch}\,\xi_1\mathrm{sh}\,\xi_1)^{\frac{1}{3}}\int_{AB}\frac{ds}{\rho^{2/3}},$$

$$\frac{\partial \omega_2}{\partial \xi_1} = \delta\int_{ABM_0}\sqrt{\dot{x}_1^2 + \dot{x}_2^2}\,d\sigma = \int_{AB}\sum_{i=1}^{2}\left(-\frac{d}{d\sigma}\frac{\partial}{\partial \dot{x}_i}\sqrt{\dot{x}_1^2 + \dot{x}_2^2}\right)\delta x_i\, d\sigma = (c^2\mathrm{ch}\,\xi_1\mathrm{sh}\,\xi_1)^{\frac{1}{3}}\int_{AB}\frac{ds}{\rho^{2/3}}, \tag{7}$$

$$\delta x_1 = \frac{\partial}{\partial \xi_1}c\,\mathrm{ch}\,\xi_1\cos\eta = c\,\mathrm{sh}\,\xi_1\cos\eta,$$

$$\delta x_2 = \frac{\partial}{\partial \xi_1}c\,\mathrm{sh}\,\xi_1\cos\eta = c\,\mathrm{ch}\,\xi_1\sin\eta$$

(here ds is an element of arclength, and ρ is the radius of curvature). The integrals over the segments MB and M_0B vanish, because the straight lines are extremals of the functional $\int\sqrt{\dot{x}_1^2 + \dot{x}_2^2}\,d\sigma$. Equation (6) is a straightforward consequence of relations (7).

It is important in carrying out the method of steepest descents to compute the phase function $\omega(\tau)$ at the saddle point. It follows from the geometric sense of ω_i $(i=1, 2)$ and Eq. (3) that the phase function $\omega(\tau)$ is equal to the length of the segment $|MM_0|$ at the saddle point (see Fig. 1).

LITERATURE CITED

1. Kazarinoff, N. D., and Ritt, R. K., Scalar diffraction by elliptic cylinder, IRE Trans. Antennas and Propagation, Suppl., AP-7 (1959).
2. Vainshtein, L. A., and Fedorov, A. A., Scattering of plane and cylindrical waves by an elliptic cylinder and the concept of diffraction rays, Radiotekh. i Elektron., 6(1):31-46 (1961).
3. Andronov, V. D., Estimates of the Greens function for the Helmholtz equation, Zh. Vychis. Matem. i Matem. Fiz., Vol. 5, No. 6 (1965).
4. Fok, V. A., Diffraction of Radio Waves around the Earth's Surface, Izd. AN SSSR (1946).
5. Friedlander, F. G., Sound Pulses, Cambridge University Press (1958).

PROPAGATION OF TRANSVERSE ELASTIC OSCILLATIONS IN THE QUANTIZED-THICKNESS WAVE-FILM MODE

V. M. Babich and T. S. Kravtsova

Consider an isotropic elastic space in which the Lamé parameters λ and μ and the density ρ of the medium have derivatives of all orders.

Let the displacement vector $\vec{u}$ have the following form in the vicinity of a certain smooth surface S:

$$\vec{u} = e^{-i\omega[t-\tau(M)]}\vec{\varphi}(M), \quad \vec{\varphi} = \{\varphi^j\}, \quad (j=1,2,3), \tag{1}$$

where ω is the frequency of the wave process and is the large parameter of the problem, t is the time, and $\tau(M)$ is a certain function of a point M in the elastic space.

We assume with regard to the functions $\varphi^j(M)$ that (1) $\varphi^j(M)$ is significantly different from zero only near S and slowly varies relative to the exponential factor; (2) the domain in which $\varphi^j(M) \sim O(1)$ has the form

$$|\nu| = O(\omega^{-1/2}),$$

and in this domain

$$\frac{\partial^{k+l+m}}{\partial s_1^k \partial s_2^l \partial \nu^m} \varphi^j(M) = O(\omega^{m/2}) \tag{2}$$

(ν is the distance to the point along the normal from S, and $\frac{\partial}{\partial s_1}$, $\frac{\partial}{\partial s_2}$ denote differentiation in the direction tangential to S).

On S we introduce a coordinate grid consisting of the curves $\tau(M) = \text{const}$ and, tangential to them, the curves $\alpha(M) = \text{const}$. In the vicinity of S the coordinates of any point of the elastic space are determined by the formula

$$\vec{X} = \vec{X}(\alpha,\tau) + \nu\vec{n}(\alpha,\tau). \tag{3}$$

Here $\vec{n}(\alpha,\tau)$ is the unit normal to S at the point (α,τ), and ν is the distance to the point $M(x,y,z)$ along the normal from S (we assume that ν has different signs in opposite directions from S), and $\vec{X} = \vec{X}(\alpha,\tau)$ is the parametric representation of the surface S.

This coordinate system, in general, is not orthogonal.

An element of length in these coordinates is expressed by the relation

$$ds = \sqrt{G_{ij}\, dq^i dq^j} \quad ,$$

in which $i, j = \alpha, \tau, \nu$, and the elements of the matrix $\| G_{ij} \|$ have the form

$$\begin{aligned}
&G_{\alpha\alpha} = |\vec{X}_\alpha|^2 + 2\nu(\vec{X}_\alpha, \vec{n}_\alpha) + |\vec{n}_\alpha|^2 \nu^2 \\
&G_{\alpha\tau} = G_{\tau\alpha} = 2(\vec{X}_\tau, \vec{n}_\alpha)\nu + (\vec{n}_\tau, \vec{n}_\alpha)\nu^2 \\
&G_{\tau\tau} = |\vec{X}_\tau|^2 + 2(\vec{X}_\tau, \vec{n}_\tau)\nu + \vec{n}_\tau^{\,2} \nu^2 \\
&G_{\alpha\nu} = G_{\nu\alpha} = G_{\tau\nu} = C_{\nu\tau} = 0 \\
&C_{\nu\nu} = 1.
\end{aligned} \tag{4}$$

The wave process is described by the dynamical equations of elasticity theory, which have the following form in an arbitrary curvilinear coordinate system:

$$\frac{\sqrt{G}}{2} \sigma_{ij} \frac{\partial G^{ij}}{\partial q^s} + \frac{\partial}{\partial q^i}\left(\sigma_{js} G^{ij} \sqrt{G}\right) - \rho\, G_{is} \sqrt{G} \frac{\partial^2 \varphi^i}{\partial t^2} = 0, \quad s = 1, 2, 3. \tag{5}$$

The iterated indices indicate summation from 1 to 3. Here $G = \det \| G_{ij} \|$, $\| G^{ij} \| = \| G_{ij} \|^{-1}$, and the components of the stress tensor σ_{ij} are related to the displacement vector by Hooke's law:

$$\sigma_{ij} = \frac{\lambda}{\sqrt{G}} \frac{\partial}{\partial q^s}\left(\varphi^s \sqrt{G}\right) G_{ij} + \mu \left[\frac{\partial G_{ij}}{\partial q^s} \varphi^s + G_{si} \frac{\partial \varphi^s}{\partial q^j} + G_{js} \frac{d\varphi^s}{\partial q^i} \right]. \tag{6}$$

We denote by φ^j in Eqs. (5) and (6) the contravariant coordinates of the displacement vector $\vec{u}$. The problem of finding solutions concentrated in the vicinity of S calls for the determination of a displacement vector $\vec{u}$ whose components have properties 1) and 2), the unknown function $\tau(M)$, and the surface S in the vicinity of which the solution is concentrated.

As it turns out, there are two distinct types of solutions in this category. One type has a transverse character, i.e., it propagates with a velocity $b(M) = \sqrt{\frac{\mu}{\rho}}$, while the other is longitudinal and has a velocity $a(M) = \sqrt{\frac{\lambda + 2\mu}{\rho}}$. In the present article we investigate transverse elastic oscillations. In [1] we have deduced the principal term of the asymptotic representation of the solution of a problem of this type; however, there are several errors in that paper, and we propose now to rederive the equations for the principal term.

§ 1. Derivation of Recursion Relations for the Components of the Displacement Vector

Let the wave process be described by Eqs. (5) and (6) with the variable velocity $b(M) = \sqrt{\frac{\mu}{\rho}}$.

We seek the components of the displacement vector in series form on reciprocal powers of the large parameter:

$$\varphi^j = \sum_{k=0}^{\infty} V_k^j(\alpha, \tau, \eta)\, \omega^{-k/2}, \tag{1.1}$$

where $\eta = \sqrt{\omega}\,\nu$, $\quad j = \alpha, \tau, \nu$.

We expand the coefficients of Eqs. (5) in power series in η, substituting the expressions for $\varphi^{j}(M)$ from Eqs. (1.1) into these equations and equating coefficients of like powers of ω.

For the leading terms (of order ω^2 and $\omega^{3/2}$) on the α and η axes we obtain the eikonal equation

$$\frac{1}{b^2}\Big|_{\nu=0} - G^{\tau\tau}\Big|_{\nu=0} = 0 \quad (1.2)$$

and the equation

$$-\frac{1}{2}\frac{\partial}{\partial\nu}\left(\frac{1}{b^2}\right)\Big|_{\nu=0} - \frac{(\vec{X}_\tau, \vec{n}_\tau)^2}{|\vec{X}_\alpha|^2 |\vec{X}_\tau|^4} = 0.$$

It follows from these equations that the derivative of $\tau(M)$ with respect to the arclength of the line $\alpha = \mathrm{const}$ is equal to $\frac{1}{b(M)}$ on the surface S and that the lines $\alpha = \mathrm{const}$ are extremals of the Fermi integral $\int \frac{ds}{b}$, i.e., they are rays. Consequently, S consists of rays. The first two coefficients of the expansion of the longitudinal component $\varphi^{\tau}(M)$ are obtained for the same powers of ω along the τ axis:

$$V_0^{\tau} = 0, \qquad V_1^{\tau} = i\frac{\partial V_0^{\nu}}{\partial\eta}. \quad (1.3)$$

Moreover, for $\omega^{-s/2}$ we obtain the system

$$\sum_{l=-2}^{s} B_l^{(i)} V_{s-l}^{\alpha} + \sum_{l=-2}^{s} D_l^{(i)} V_{s-l}^{\tau} + \sum_{l=-2}^{s} K_l^{(i)} V_{s-l}^{\nu} = 0, \quad (1.4)$$

$$s = -2, -1, 0, 1, \ldots, \qquad i = \alpha, \tau, \nu.$$

Here $B_l^{(i)}, D_l^{(i)}, K_l^{(i)}$ denote second-order linear differential operators, the coefficients of which are known polynomials in the variable η. We shall omit the cumbersome explicit expressions for these coefficients.

We obtain the following from (1.4) for the coefficients of the expansion of the longitudinal component $\varphi^{\tau}(M)$:

$$V_{s+2}^{\tau} = i\frac{\partial V_{s+1}^{\nu}}{\partial\eta} + \frac{1}{\lambda+\mu}\left[\sum_{l=-2}^{s} B_l^{(i)} V_{s-l}^{\alpha} + \sum_{l=-2}^{s} D_l^{(i)} V_{s-l}^{\tau} + \sum_{l=-2}^{s} K_l^{(i)} V_{s-l}^{\nu}\right], \quad (1.5)$$

$$s = -1, 0, 1, 2, \ldots, \qquad i = \alpha, \tau, \nu.$$

We insert the value found for V_{s+2}^{τ} into the other equations of system (1.4) and obtain a system of two second-order differential equations for the coefficients of the transverse components φ^{α} and φ^{ν}:

$$\left.\begin{aligned} L(|\vec{X}_\alpha| V_{s+2}^{\alpha}) + 2i\frac{(\vec{X}_\alpha, \vec{n}_\tau)}{b^2|\vec{X}_\alpha|} V_{s+2}^{\nu} &= f_s^{\alpha}, \\ L(V_{s+2}^{\nu}) + 2i\frac{(\vec{X}_\alpha, \vec{n}_\tau)}{b^2} V_{s+2}^{\alpha} &= f_s^{\nu}, \end{aligned}\right\} \quad (1.6)$$

where

$$L = \frac{\partial^2}{\partial \eta^2} + \frac{2i}{b^2}\frac{\partial}{\partial \tau} + \frac{i}{b^2}\frac{\partial}{\partial \tau}\ln\frac{\mu|\vec{X}_\alpha|}{b} - \left[\frac{b_{\nu\nu}}{b^3}\bigg|_{\nu=0} + \frac{3(\vec{X}_\alpha,\vec{n}_\tau)^2}{b^4|\vec{X}_\alpha|^2}\right] \tag{1.7}$$

and the right-hand sides have the form

$$\begin{aligned} f_s^\alpha &= -\frac{1}{\mu|\vec{X}_\alpha|}\left[\sum_{l=-1}^{s} B_l^{(\alpha)} V_{s-l}^\alpha + \sum_{l=-1}^{s} D_l^{(\alpha)} V_{s-l+1}^\tau + \sum_{l=-1}^{s} K_l^{(\alpha)} V_{s-l}^\nu\right], \\ f_s^\nu &= -\frac{1}{\mu}\left[\sum_{l=-1}^{s} B_l^{(\nu)} V_{s-l}^\alpha + \sum_{l=-1}^{s} D_l^{(\nu)} V_{s-l+1}^\tau + \sum_{l=-1}^{s} K_l^{(\nu)} V_{s-l}^\nu\right]. \end{aligned} \tag{1.8}$$

The system of equations (1.6) is recursive and can be used to successively determine all terms of the expansion of the transverse components φ^α and φ^ν up to terms of order $\omega^{-N/2}$, where N is any predetermined integer. The subcomponents of the longitudinal component φ^τ are expressed in terms of the subcomponents of the transverse components φ^α and φ^ν and their derivatives according to Eqs. (1.5).

§ 2. Zeroth Approximation for φ^α and φ^ν

We define the zeroth approximation as the principal terms V_0^α and V_0^ν of the expansions (1.1). Their equations are deduced from Eqs. (1.6) with $s=-2$:

$$\left.\begin{aligned} L(|\vec{X}_\alpha| V_0^\alpha) + 2i\frac{(\vec{X}_\alpha,\vec{n}_\tau)}{b^2|\vec{X}_\alpha|}V_0^\nu &= 0, \\ L(V_0^\nu) + 2i\frac{(\vec{X}_\alpha,\vec{n}_\tau)}{b^2}V_0^\alpha &= 0. \end{aligned}\right\} \tag{2.1}$$

The operator L has the form (1.7).

We first investigate the equation

$$L W_0 = 0.$$

It is entirely analogous to the equation investigated in [2]. Proceeding in similar fashion, we put

$$\varsigma = \psi(\alpha,\tau)\eta \qquad (\eta = \sqrt{\omega}\,\nu),$$

$$W_0 = \sqrt{\psi}\exp\left[-\frac{i\psi_\tau}{2b^2\psi^3}\varsigma^2\right]W,$$

where $\psi(\alpha,\tau)$ is for the time being an unknown function of α and τ.

Then the function W must satisfy the equation

$$W_{\varsigma\varsigma} + \varsigma^2\left[\left(\frac{\psi_\tau}{b^2\psi^3}\right)^2 + \frac{1}{b^2\psi^2}\frac{\partial}{\partial\tau}\left(\frac{\psi_\tau}{b^2\psi^3}\right) - \frac{1}{\psi^4}\left(\frac{b_{\nu\nu}}{b^3}\bigg|_{\nu=0} + \frac{3(\vec{X}_\alpha,\vec{n}_\tau)^2}{|\vec{X}_\alpha|^2 b^4}\right)\right]W + \frac{2i}{b^2\psi^3}W_\tau + \frac{1}{b^2\psi^2}\frac{\partial}{\partial\tau}\ln\frac{\mu|\vec{X}_\alpha|}{b}W = 0. \tag{2.2}$$

If the expression in the brackets is constant and equal to C, then separation of variables is possible in the latter equation. It is may be assumed without loss of generality that $C=-1$,* i.e., that

*When $C>0$, the solution is not concentrated in the vicinity of S ($\nu=0$), and in the case $C<0$, $C\neq-1$ it cancels out in the final equations.

$$\left(\frac{\Psi_\tau}{b^2\Psi^3}\right)^2 + \frac{1}{b^2\Psi^2}\frac{\partial}{\partial\tau}\left(\frac{\Psi_\tau}{b^2\Psi^3}\right) - \frac{1}{\Psi^4}\left[\frac{b_{\nu\nu}}{b^3}\Big|_{\nu=0} + \frac{3(\vec{X}_\alpha, \vec{n}_\tau)^2}{|\vec{X}_\alpha|^2 b^4}\right] = -1. \tag{2.3}$$

The unknown function $\Psi(\alpha,\tau)$ is now determined from Eq. (2.3). The substitution $\Psi(\alpha,\tau) = \frac{1}{b\,\beta(\alpha,\tau)}$ reduces Eq. (2.2) to the form

$$\beta'' + \mathcal{F}(\alpha,\tau)\beta = \frac{1}{\beta^3} \quad \mathcal{F}(\alpha,\tau) = \frac{b_{\tau\tau}}{b} - 2\frac{b_\tau^2}{b^2} - b\,b_{\nu\nu}\Big|_{\nu=0} - \frac{3(\vec{X}_\alpha, \vec{n}_\tau)^2}{|\vec{X}_\alpha|^2}. \tag{2.4}$$

We know that the general solution of the latter equation (see, e.g., [2]) is determined by the equation

$$\beta(\tau) = \left[\sum_{j,k=1}^{2} d_{jk}\, y_j(\tau)\, y_k(\tau)\right]^{1/2}, \tag{2.5}$$

where y_1, y_2 is any pair of linearly independent solutions of the equation

$$y'' + \mathcal{F}(\alpha,\tau)\, y = 0$$

and the determinant of the positive-definite matrix $\|d_{jk}\|$ satisfies the relation

$$\det\|d_{jk}\| \cdot (y_1 y_2' - y_1' y_2)^2 = 1.$$

We shall assume hereinafter that the function $\Psi(\alpha,\tau)$ is known.

Given condition (2.3), the variables in Eq. (2.2) are separable, and the function W becomes equal to

$$W = \sqrt{\frac{b}{|\vec{X}_\alpha|}} \exp\left[-\frac{im}{2}\int b^2\Psi^2 d\tau\right] Z(\zeta),$$

where m is the separation constant and the function $Z(\zeta)$ is subject to the Weber equation

$$Z'' + (m - \zeta^2) Z = 0.$$

Only for $m = 2n+1$, $n = 0, 1, 2, \ldots$, does the Weber equation have solutions that tend to zero at infinity. In this case

$$Z = Z_n(\zeta) = e^{-\frac{\zeta^2}{2}} H_n(\zeta),$$

where $H_n(\zeta)$ is the nth Hermite polynomial.

For $|\zeta| \leq \sqrt{2n+1}$ the functions $Z_n(\zeta)$ oscillate, and for $|\zeta| > \sqrt{2n+1}$ they tend monotonically to zero. Consequently, the thickness of the layer spanned by the oscillations is equal to

$$|\zeta| = \sqrt{2n+1}, \quad n = 0, 1, 2, \ldots, \quad \left(|\nu| = \frac{\omega^{-1/2}}{\Psi}\sqrt{2n+1}\right).$$

The final equation for W_0 has the form

$$W_0(\alpha,\tau,\nu) = W_{0,n} = \sqrt{\frac{\Psi}{\beta^4}}\sqrt{\frac{b}{|\vec{X}_\alpha|}} \exp\left[-\frac{i(2n+1)}{2}\int b^2\Psi^2 d\tau - \frac{\zeta^2}{2}\left(1 + \frac{i\Psi_\tau}{b^2\Psi^3}\right)\right] H_n(\zeta), \tag{2.6}$$

where $\zeta = \Psi(\alpha,\tau)\eta$ $(\eta = \sqrt{\omega}\,\nu)$, $\Psi(\alpha,\tau) = \frac{1}{b\,\beta(\alpha,\tau)}$, and the function $\beta(\alpha,\tau)$ is determined by Eq. (2.5).

It follows from the foregoing that the solution of the zeroth approximation (2.1) has the form

$$V_0^\alpha = \frac{1}{|\vec{X}_\alpha|}\Phi_0^\alpha(\tau)W_0 \qquad (2.7)$$
$$V_0^\nu = \Phi_0^\nu(\tau)W_0,$$

where the functions $\Phi_0^\alpha(\tau)$ and $\Phi_0^\nu(\tau)$ satisfy the system of ordinary first-order equations

$$\left.\begin{aligned} &\frac{\partial\Phi_0^\alpha}{\partial\tau} + \frac{(\vec{X}_\alpha, \vec{n}_\tau)}{|\vec{X}_\alpha|}\Phi_0^\nu = 0 \\ &\frac{\partial\Phi_0^\nu}{\partial\tau} + \frac{(\vec{X}_\alpha, \vec{n}_\tau)}{|\vec{X}_\alpha|}\Phi_0^\alpha = 0. \end{aligned}\right\} \qquad (2.8)$$

Adding both equations of the system, we obtain

$$\frac{\partial(\Phi_0^\alpha + \Phi_0^\nu)}{\partial\tau} + \frac{(\vec{X}_\alpha, \vec{n}_\tau)}{|\vec{X}_\alpha|}(\Phi_0^\alpha + \Phi_0^\nu) = 0,$$

whence

$$\Phi_0^\alpha + \Phi_0^\nu = 2C_1(\alpha)\exp\left[-\int\frac{(\vec{X}_\alpha, \vec{n}_\tau)}{|\vec{X}_\alpha|}d\tau\right].$$

Subtracting the second equation of (2.8) from the first, we obtain the difference between the functions we are seeking:

$$\Phi_0^\alpha - \Phi_0^\nu = 2C_2(\alpha)\exp\left[\int\frac{(\vec{X}_\alpha, \vec{n}_\tau)}{|\vec{X}_\alpha|}d\tau\right].$$

Consequently,

$$\Phi_0^\alpha(\tau) = C_1(\alpha)\exp\left[-\int\frac{(\vec{X}_\alpha, \vec{n}_\tau)}{|\vec{X}_\alpha|}d\tau\right] + C_2(\alpha)\exp\left[\int\frac{(\vec{X}_\alpha, \vec{n}_\tau)}{|\vec{X}_\alpha|}d\tau\right] \qquad (2.9)$$
$$\Phi_0^\nu(\tau) = C_1(\alpha)\exp\left[-\int\frac{(\vec{X}_\alpha, \vec{n}_\tau)}{|\vec{X}_\alpha|}d\tau\right] - C_2(\alpha)\exp\left[\int\frac{(\vec{X}_\alpha, \vec{n}_\tau)}{|\vec{X}_\alpha|}d\tau\right].$$

Therefore, the solution of the zeroth-approximation system has the following form up to an arbitrary function α:

$$V_0^\alpha = V_{0,n}^\alpha = \frac{1}{|\vec{X}_\alpha|}\sqrt{\frac{\Psi}{\rho^4}}\sqrt{\frac{\sigma}{|\vec{X}_\alpha|}}\exp\left[-\frac{i(2n+1)}{2}\int\sigma^2\psi^2 d\tau - \frac{\xi^2}{2}\left(1 + \frac{i\psi_\tau}{\sigma^2\psi^3}\right)\right]H_n(\xi) \qquad (2.10)$$

$$V_0^\nu = V_{0,n}^\nu = \sqrt{\frac{\Psi}{\rho^4}}\sqrt{\frac{\sigma}{|\vec{X}_\alpha|}}\exp\left[-\frac{i(2n+1)}{2}\int\sigma^2\psi^2 d\tau - \frac{\xi^2}{2}\left(1 + \frac{i\psi_\tau}{\sigma^2\psi^3}\right)\right]H_n(\xi),$$

$$n = 0, 1, 2, \ldots,$$

where $\Phi_0^\kappa(\tau)$ $(\kappa = \alpha, \nu)$ is determined by Eqs. (2.9).

§ 3. Higher Approximations for φ^α and φ^ν

The higher approximations are determined from the system (1.6) with $s = -1, 0, 1, 2, \ldots$ Using the equation for the Hermite polynomials:

$$\mathcal{H}_n''(\zeta) - 2\zeta\,\mathcal{H}_n'(\zeta) + 2n\,\mathcal{H}_n(\zeta) = 0$$

and the recursion relations

$$\zeta\mathcal{H}_n(\zeta) = \frac{1}{2}\mathcal{H}_{n+1}(\zeta) + n\,\mathcal{H}_{n-1}(\zeta), \quad \mathcal{H}_n'(\zeta) = 2n\,\mathcal{H}_{n-1}(\zeta),$$

we readily verify that the right-hand sides of the system (2.6) have the form

$$f_s^j = \sqrt{\frac{\Psi}{\rho^\mu}}\sqrt{\frac{6}{|\vec{X}_\alpha|}}\exp\left[-\frac{i(2n+1)}{2}\int 6^2\psi^2 d\tau - \frac{\zeta^2}{2}\left(1+\frac{i\psi_\tau}{6^2\psi^3}\right)\right]\sum_{\kappa\le n+s+4} p^j_{s,\kappa}(\alpha,\tau)\,\mathcal{H}_\kappa(\zeta), \tag{3.1}$$

where

$$j = \alpha, \nu, \qquad \kappa = \begin{cases} 1,3,5,\ldots & \text{for even } n; \\ 0,2,4,\ldots & \text{for odd } n, \end{cases}$$

and $p^j_{s,\kappa}(\alpha,\tau)$ are known functions of α and τ. We seek the solution of this system in the same form, but with indeterminate coefficients $a^\alpha_{s+2,\kappa}(\alpha,\tau)$ and $a^\nu_{s+2,\kappa}(\alpha,\tau)$, i.e.,

$$V^\alpha_{s+2,\kappa} = \frac{1}{|\vec{X}_\alpha|}\sqrt{\frac{\Psi}{\rho^\mu}}\sqrt{\frac{6}{|\vec{X}_\alpha|}}\exp\left[-\frac{i(2n+1)}{2}\int 6^2\psi^2 d\tau - \frac{\zeta^2}{2}\left(1+\frac{i\psi_\tau}{6^2\psi^3}\right)\right]\sum_{\kappa\le n+s+4} a^\alpha_{s+2,\kappa}(\alpha,\tau)\,\mathcal{H}_\kappa(\zeta) \tag{3.2}$$

$$V^\nu_{s+2,\kappa} = \sqrt{\frac{\Psi}{\rho^\mu}}\sqrt{\frac{6}{|\vec{X}_\alpha|}}\exp\left[-\frac{i(2n+1)}{2}\int 6^2\psi^2 d\tau - \frac{\zeta^2}{2}\left(1+\frac{i\psi_\tau}{6^2\psi^3}\right)\right]\sum_{\kappa\le n+s+4} a^\nu_{s+2,\kappa}(\alpha,\tau)\,\mathcal{H}_\kappa(\zeta), \tag{3.3}$$

where

$$\kappa = \begin{cases} 1,3,5,\ldots & \text{for even } n; \\ 0,2,4,\ldots & \text{for odd } n. \end{cases}$$

Applying the operator on the left-hand side of the system (1.6) to these functions, we arrive once again at a linear combination of Hermite polynomials of the type (3.1). Equating coefficients of polynomials of like orders (they are linearly independent for integer-valued n), we obtain a system of two ordinary differential equations in the functions $a^\alpha_{s+2,\kappa}(\alpha,\tau)$, $a^\nu_{s+2,\kappa}(\alpha,\tau)$:

$$\left.\begin{aligned} \frac{\partial a^\alpha_{s+2,\kappa}}{\partial\tau} + \frac{(\vec{X}_\alpha,\vec{n}_\tau)}{|\vec{X}_\alpha|}a^\nu_{s+2,\kappa} &= -\frac{i6^2}{2}p^\alpha_{s+2,\kappa} \\ \frac{\partial a^\nu_{s+2,\kappa}}{\partial\tau} + \frac{(\vec{X}_\alpha,\vec{n}_\tau)}{|\vec{X}_\alpha|}a^\alpha_{s+2,\kappa} &= -\frac{i6^2}{2}p^\nu_{s+2,\kappa}, \end{aligned}\right\} \tag{3.4}$$

in which $p^j_{s+2,\kappa}(\alpha,\tau)$ are known functions of α and τ, $s = -1,0,1,\ldots$. Solving this system, we obtain, up to an arbitrary function of α,

$$a^{j}_{s+2,\kappa} = -\frac{i}{4}\left[e^{-\int \frac{(\vec{X}_\alpha,\vec{n}_\tau)}{|\vec{X}_\alpha|}d\tau}\int 6^2\left(p^{\alpha}_{s+2,\kappa}+p^{\nu}_{s+2,\kappa}\right)e^{\int \frac{(\vec{X}_\alpha,\vec{n}_\tau)}{|\vec{X}_\alpha|}d\tau} \pm e^{\int \frac{(\vec{X}_\alpha,\vec{n}_\tau)}{|\vec{X}_\alpha|}d\tau}\int 6^2\left(p^{\alpha}_{s+2,\kappa}-p^{\nu}_{s+2,\kappa}\right)e^{-\int \frac{(\vec{X}_\alpha,\vec{n}_\tau)}{|\vec{X}_\alpha|}d\tau}\right],$$

where $j=\alpha,\nu$, the plus sign is chosen for $a^{\alpha}_{s+2,\kappa}$, the minus sign is chosen for $a^{\nu}_{s+2,\kappa}$, and $s=-1,0,1,\ldots$

§ 4. Conclusion

Consolidating the equations obtained above, we infer the following asymptotic representation (as $\omega \to \infty$) of the displacement vector $\vec{u}$:

$$\vec{u} = e^{-i\omega[t-\tau(M)]}\left(|\vec{X}_\alpha| V^{\alpha}_{0},\ \omega^{-1/2}|\vec{X}_\tau| V^{\tau}_{1},\ V^{\nu}_{0}\right), \tag{4.1}$$

where

$$V^{\alpha}_{0} = \frac{1}{|\vec{X}_\alpha|} W_0 \Phi^{\alpha}_{0},\quad V^{\nu}_{0} = W_0 \Phi^{\nu}_{0},\quad V^{\tau}_{1} = i\frac{\partial V^{\nu}_{0}}{\partial \eta},$$

$$W_0 = \sqrt{\frac{\psi}{\rho}}\sqrt{\frac{6}{|\vec{X}_\alpha|}}\exp\left[-\frac{i(2n+1)}{2}\int 6^2\psi^2 d\tau - \frac{i\psi_\tau}{6^2\psi^3}\varsigma^2\right]e^{-\frac{\varsigma^2}{2}}H_n(\varsigma), \tag{4.2}$$

$$n=0,1,2,\ldots,$$

where $\Phi^{j}_{0}\ (j=\alpha,\nu)$ are determined by Eqs. (2.9), $\varsigma = \psi\eta = \sqrt{\omega}\,\psi\nu$, $\psi(\alpha,\tau) = \frac{1}{6\beta(\alpha,\tau)}$ is a positive function, $\beta(\alpha,\tau)$ is determined by Eqs. (2.4)-(2.5), and

$$\tau(M) = \int_{M_0}^{M}\frac{ds}{6(M)}. \tag{4.3}$$

As noted in § 2, the form of the function W_0 implies that the solution (4.1) is concentrated in the vicinity of the surface $S(\nu=0)$ in strips of thickness

$$|\nu| = \frac{\omega^{-1/2}}{\psi(\alpha,\tau)}\sqrt{2n+1},\quad n=0,1,2,\ldots, \tag{4.4}$$

and we therefore call solutions of the type (4.1) quantized-thickness wave films. It is readily perceived that the main particle displacement of the medium is parallel to the plane (α,ν), which is perpendicular to the direction of wave propagation, i.e., the solutions have a transverse character. It can be shown that waves of the type (4.1) satisfy the energy conservation law in the sense that the energy contained in a certain small volume $d\Omega$ does not change as this volume moves along the ray-extremals of the integral $\int\frac{ds}{6(M)}$ with the wave velocity $6(M)=\sqrt{\frac{\mu}{\rho}}$.

Thus, if the wave process is described by Eqs. (5)-(6), the energy density has the form

$$\frac{\rho}{2}\vec{u}_t\vec{u}^{*}_t + \frac{1}{2}\sigma_{i'j'}(\vec{u})\,\varepsilon_{ij}(\vec{u}^{*})\,G^{i'i}G^{j'j}. \tag{4.5}$$

We recall that σ_{ij} are the components of the stress tensor, ε_{ij} are the components of the strain tensor, and

$$\sigma_{ij} = \lambda\,\mathrm{div}\,\vec{u}\,G_{ij} + 2\mu\,\varepsilon_{ij}(\vec{u}).$$

The principal energy term dE inside a small volume

$$\alpha_0 \leq \alpha \leq \alpha_0 + d\alpha, \quad t \leq \tau \leq t + d\tau, \quad -\nu_0 \leq \nu \leq \nu_0, \quad \nu_0 = O(1) \tag{4.6}$$

has the form

$$dE = \omega^2 \psi^2 [|\Psi_0^\alpha|^2 + |\Phi_0^\nu|^2] \, d\tau \, d\alpha \int_{-\nu_0}^{\nu_0} [e^{-\frac{\zeta^2}{2}} H_n(\zeta)]^2 d\zeta.$$

Relying on the fact that $\zeta = \sqrt{\omega} \, \psi(\alpha, \tau) \, \nu$ and replacing the integration limits by $\pm\infty$, we arrive at the equation

$$dE = \omega^{3/2} [|\Phi_0^\alpha|^2 + |\Phi_0^\nu|^2] \sqrt{2\pi} \, n! \, d\tau \, d\alpha. \tag{4.7}$$

This expression depends only on the ray, i.e., the principal energy term in the small volume (4.6) is independent of the time.

LITERATURE CITED

1. Babich, V. M., and Kravtsova, T. S., Prikl. Matem. i Mekh., 31(2):204-210 (1967).
2. Babich, V. M., and Lazutkin, V. F., Eigenfunctions concentrated near a closed geodesic, Topics in Mathematical Physics, Vol. 2, Consultants Bureau, New York (1968), pp. 9-18.
3. Whittaker, E. T., and Watson, G. N., Modern Analysis, Cambridge University Press (1927).

DIFFRACTION OF A CYLINDRICAL WAVE BY A SEMI-INFINITE PLATE AT THE BOUNDARY BETWEEN TWO ACOUSTIC MEDIA

B. P. Belinskii, D. P. Kouzov,
V. D. Luk'yanov, and V. D. Chel'tsova

In this article we investigate the steady-state oscillations of two adjoining acoustic media. A semi-infinite baffle is situated at their interface, which is assumed to be rectilinear. We shall formulate a solution to the diffraction problem at the edge of the baffle for a cylindrical wave generated by a "filamentary" source located in one of the media. The axis of the cylindrical wave is presumably parallel to the edge of the plate (two-dimensional problem).

We regard the baffle as a perfectly rigid plate (Problem 1) or a plate free to oscillate in flexural modes (Problem 2). In the latter case, therefore, we disregard longitudinal motion of the plate. This approximation is valid when the acoustical properties of the media are sufficiently different [1], a situation that we shall take for granted henceforth. Allowance for symmetrical motions of the plate, which contribute significantly to the field in the event that the acoustical properties of the media are identical or very similar [2-4], would complicate the problem in the fundamental mathematical respect (Riemann matrix problem). We shall also assume the plate is of infinitesimal thickness. In other words, we disregard any phase losses suffered in wave transmission through the plate, as well as wave reflection from the cross-sectional plane of the plate ("end effect"). Thus, the plate thickness does not influence the geometry of the problem, but is merely taken into account as a certain mechanical constant in specifying the acoustical properties of the plate. Under the foregoing assumptions with regard to the second problem we shall formulate a rigorous mathematical solution for both problems. We shall not analyze the solution. The analogous problem for a rigid baffle and unsteady incident plane wave as the field source has been investigated in [5].

Notation

$\rho_{1,2}$ - density of the medium	E_0 - Young's modulus of the plate
ω - oscillation frequency	h - plate thickness
$P_{1,2}$ - pressure	ρ_0 - plate density
$\kappa_{1,2}$ - wave numbers	

The time factor $e^{-i\omega t}$ is omitted everywhere.

§ 1. Statement of the Problem

The first medium is situated in the half-plane $0 < z < +\infty$, $-\infty < x < +\infty$, and the second is situated in the half-plane $-\infty < z < 0$, $-\infty < x < +\infty$. The semi-infinite baffle is located at $z=0$, $-\infty < x < 0$. The point source is located at the point $(x_0, z_0 < 0)$ (Fig. 1).

The state of the medium is characterized in terms of the pressures P_1 and P_2, which obey the Helmholtz equations in the corresponding half-planes: $(\Delta + \kappa_1^2) P_1 = 0$,

$$(\Delta + \kappa_2^2) P_2 = \delta(x - x_0; z - z_0). \tag{1.1}$$

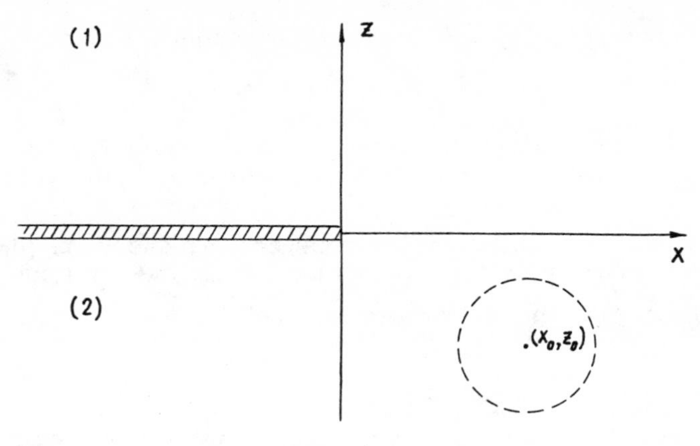

Fig. 1

The pressures also satisfy appropriate boundary conditions governing the acoustic coupling between the media. On the positive side of the axis OX, where the two media are in direct contact, the pressures are continuous:

$$P_1 = P_2, \qquad (z = 0, \quad 0 < x < \infty), \tag{1.2}$$

as are the normal displacements:

$$\frac{1}{\rho_1}\frac{\partial P_1}{\partial z} = \frac{1}{\rho_2}\frac{\partial P_2}{\partial z}, \quad (z = 0, \quad 0 < x < \infty). \tag{1.3}$$

On the negative side of the axis OX, where the media are separated by the plate, zero normal displacements are assumed in the case of a rigid plate:

$$\frac{1}{\rho_1}\frac{\partial P_1}{\partial z} = \frac{1}{\rho_2}\frac{\partial P_2}{\partial z} = 0, \quad (z = 0, -\infty < x < 0). \tag{1.4}$$

The case of an elastic plate is governed by the equations

$$\frac{1}{\rho_1}\frac{\partial P_1}{\partial z} = \frac{1}{\rho_2}\frac{\partial P_2}{\partial z}, \tag{1.5}$$

$$P_1 - P_2 = m\left(-i\frac{\partial}{\partial x}\right)\left(\frac{1}{\rho_1}\frac{\partial P_1}{\partial z} + \frac{1}{\rho_2}\frac{\partial P_2}{\partial z}\right), \quad (z = 0, \ -\infty < x < 0). \tag{1.6}$$

The second equation connotes equilibrium of the forces acting on the plate. Here we have assumed

$$m\left(-i\frac{\partial}{\partial x}\right) = h\rho_0 - \frac{E_0 h^3}{3\omega^2}\left(-i\frac{\partial}{\partial x}\right)^4. \tag{1.7}$$

We note that Eq. (1.3) holds in both problems (1 and 2) over the entire axis OX, a result that is essential to the ensuing computations. Allowance for symmetrical motions of the plate would violate relation (1.5).

Also, in the case of an elastic plate it is necessary to formulate its end conditions. We require here that the point moment and point force vanish:

$$\begin{gathered}\frac{\partial^2}{\partial x^2}\left(\frac{1}{\rho_1}\frac{\partial P_1}{\partial z} + \frac{1}{\rho_2}\frac{\partial P_2}{\partial z}\right) = 0, \\ \frac{\partial^3}{\partial x^3}\left(\frac{1}{\rho_1}\frac{\partial P_1}{\partial z} + \frac{1}{\rho_2}\frac{\partial P_2}{\partial z}\right) = 0, \qquad (z = 0, \ x \to -0).\end{gathered} \tag{1.8}$$

We regard the wave numbers here and elsewhere as complex, so that

$$\mathrm{Re}(\kappa_1, \kappa_2) > 0, \qquad \mathrm{Im}(\kappa_1; \kappa_2) > 0.$$

We seek the solution in the class of functions that decay exponentially to infinity. The solution for real κ_1 and κ_2 is interpreted as the limit of the solution for complex κ_1 and κ_2 as $\mathrm{Im}(\kappa_1, \kappa_2) \to +0$ (absorption limit principle).

We formulate the solutions by the method of expansion in plane waves [6, 7], then reduce the problem to inhomogeneous boundary-value problems for the analytic functions (8).

We begin (§ 2) by finding the special representation of the field of the point source in the absence of the baffle.

§ 2. Field of the Source in the Absence of the Baffle

We consider the harmonic motions of a pair of homogeneous acoustic media due to a source concentrated at the point $x = x_0$, $z = z_0 < 0$. The pressures obey the Helmholtz equations

$$(\Delta + \kappa_1^2) P_1 = 0, \quad (z > 0, -\infty < x < +\infty), \tag{2.1}$$

$$(\Delta + \kappa_2^2) P_2 = \delta(x - x_0, z - z_0), \quad (z < 0, -\infty < x < +\infty). \tag{2.2}$$

Equations (2.1) and (2.2) must be solved under the boundary conditions

$$P_1 = P_2, \quad \frac{1}{\rho_1}\frac{\partial P_1}{\partial z} = \frac{1}{\rho_2}\frac{\partial P_2}{\partial z}, \tag{2.3}$$

which we immediately assume are fulfilled on the entire axis OX.

We seek the solution in the form

$$\begin{gathered} P_1(x, z) = \int_{-\infty}^{+\infty} \tau_1(\lambda) e^{i\lambda x - s_1(\lambda) z} \, d\lambda, \\ P_2(x, z) = \int_{-\infty}^{+\infty} \tau_2(\lambda) e^{i\lambda x + s_2(\lambda) z} \, d\lambda - \frac{1}{4\pi} \int_{-\infty}^{+\infty} \frac{1}{s_2(\lambda)} e^{i\lambda(x - x_0) - s_2(\lambda)|z - z_0|} \, d\lambda, \end{gathered} \tag{2.4}$$

in which we have set

$$s_i(\lambda) = \sqrt{\lambda^2 - \kappa_i^2}, \quad (i = 1, 2). \tag{2.5}$$

The branches of the radicals are fixed by the requirement

$$\lim_{\lambda \to \pm\infty} [\mathrm{Im}\, s_{1,2}(\lambda)] = 0, \quad (\mathrm{Im}\,\lambda = 0). \tag{2.6}$$

The branch cuts in the plane of λ are drawn between the points

$\lambda = \kappa_1$ and $\lambda = \kappa_2$, $\lambda = \kappa_1$ and $\lambda = \infty$,

$\lambda = -\kappa_1$ and $\lambda = -\kappa_2$, $\lambda = -\kappa_1$ and $\lambda = \infty$.

The cuts are centrosymmetric about the origin and do not cross the real axis (Fig. 2).

In this form the solutions allow Eqs. (2.1) and (2.2) as well as [for a very large class of functions $\tau_{1,2}(\lambda)$] the conditions at infinity to be satisfied.

The boundary conditions (2.3) yield two relations, which hold for all λ:

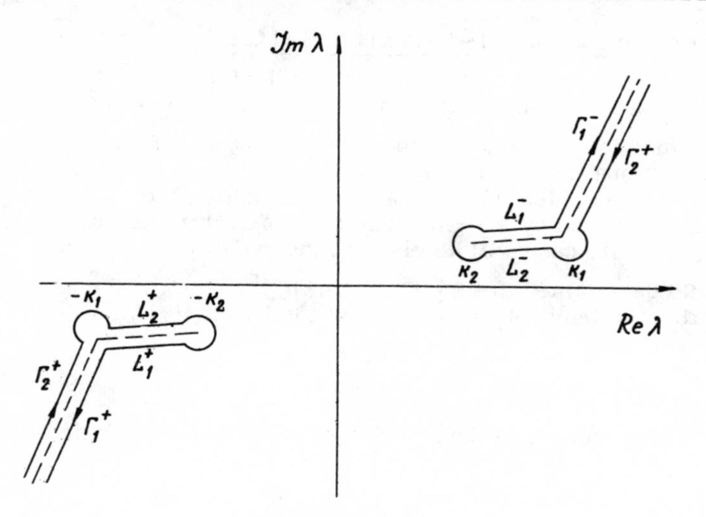

Fig. 2.

$$\int_{-\infty}^{+\infty}[\tau_1(\lambda)-\tau_2(\lambda)]e^{i\lambda x}d\lambda=-\frac{1}{4\pi}\int_{-\infty}^{\infty}\frac{1}{s_2(\lambda)}e^{i\lambda(x-x_0)+s_2(\lambda)(z_0)}d\lambda,$$

$$\int_{-\infty}^{+\infty}\left[-\frac{s_1(\lambda)}{\rho_1}\tau_1(\lambda)-\frac{s_2(\lambda)}{\rho_2}\tau_2(\lambda)\right]e^{i\lambda x}d\lambda=-\frac{1}{4\pi}\int_{-\infty}^{+\infty}\frac{1}{\rho_2}e^{i\lambda(x-x_0)+s_2(\lambda)z_0}d\lambda. \tag{2.7}$$

If (2.7) is to be satisfied, it suffices to put

$$\tau_1(\lambda)-\tau_2(\lambda)=-\frac{1}{4\pi}\cdot\frac{e^{-i\lambda x_0+s_2(\lambda)z_0}}{s_2(\lambda)},$$

$$-\frac{s_1(\lambda)}{\rho_1}\tau_1(\lambda)-\frac{s_2(\lambda)}{\rho_2}\tau_2(\lambda)=\frac{1}{4\pi}\frac{e^{-i\lambda x_0+s_2(\lambda)z_0}}{\rho_2}. \tag{2.8}$$

We have finally

$$\tau_1(\lambda)=-\frac{1}{4\pi}\frac{2\rho_1 e^{-i\lambda x_0+s_2(\lambda)z_0}}{\rho_1 s_2(\lambda)+\rho_2 s_1(\lambda)},$$

$$\tau_2(\lambda)=\frac{1}{4\pi}\cdot\frac{[\rho_2 s_1(\lambda)-\rho_1 s_2(\lambda)]e^{i\lambda x_0+s_2(\lambda)z_0}}{s_2(\lambda)[\rho_1 s_2(\lambda)+\rho_2 s_1(\lambda)]}, \tag{2.9}$$

whereupon

$$P_1(x,z)=-\frac{1}{4\pi}\int_{-\infty}^{+\infty}\frac{2\rho_1}{\rho_1 s_2(\lambda)+\rho_2 s_1(\lambda)}e^{i\lambda(x-x_0)-s_1(\lambda)z+s_2(\lambda)z_0}d\lambda,$$

$$P_2(x,z)=-\frac{1}{4\pi}\int_{-\infty}^{+\infty}\frac{1}{s_2(\lambda)}\left[\frac{\rho_1 s_2(\lambda)-\rho_2 s_1(\lambda)}{\rho_1 s_2(\lambda)+\rho_2 s_1(\lambda)}e^{s_2(\lambda)(z+z_0)}+e^{-s_2(\lambda)|z-z_0|}\right]e^{i\lambda(x-x_0)}d\lambda. \tag{2.10}$$

§ 3. Wave Incidence on a Semi-infinite Baffle

We now consider the diffraction field generated by the incidence of the wave described in § 2 on a semi-infinite baffle. The same assumptions as above are imposed on the wave numbers.

We demand of the functions P_1 and P_2 exponential decay at infinity and continuity on the plane (x, z) with a discontinuity along the negative side of the axis OX to the branch cut. We shall formulate the solution of the diffraction problem for the case of an elastic plate. The calculations pertinent to the simpler problem of diffraction by a rigid baffle are omitted in the interest of conserving space. The solution for the rigid baffle case will be given at the end of the section.

We denote the field determined in § 2 [Eq. (2.10)] by P_1°, P_2°. We now introduce the diffraction field Q_1, Q_2 separating the incident wave from the total field P_1, P_2 :

$$Q_{1,2} = P_{1,2} - P_{1,2}^\circ. \tag{3.1}$$

The diffraction field is subject to the homogeneous Helmholtz equations:

$$\begin{gathered}(\Delta + K_1^2)\, Q_1 = 0, \quad (z > 0,\ -\infty < x < +\infty),\\ (\Delta + K_2^2)\, Q_2 = 0, \quad (z < 0,\ -\infty < x < +\infty).\end{gathered} \tag{3.2}$$

The boundary equations take the form

$$Q_1 - Q_2 = 0, \quad (z = 0,\ 0 < x < +\infty), \tag{3.3}$$

$$Q_1 - Q_2 - m\left(-i\frac{\partial}{\partial x}\right)\left(\frac{1}{\rho_1}\frac{\partial Q_1}{\partial z} + \frac{1}{\rho_2}\frac{\partial Q_2}{\partial z}\right) = m\left(-i\frac{\partial}{\partial x}\right)\left(\frac{1}{\rho_1}\frac{\partial P_1^\circ}{\partial z} + \frac{1}{\rho_2}\frac{\partial P_2^\circ}{\partial z}\right), \tag{3.4}$$

$$(z = 0,\ -\infty < x < 0),$$

$$\frac{1}{\rho_1}\frac{\partial Q_1}{\partial z} - \frac{1}{\rho_2}\frac{\partial Q_2}{\partial z} = 0, \quad (z = 0,\ -\infty < x < +\infty). \tag{3.5}$$

The following contact conditions must also hold:

$$\frac{\partial^2}{\partial x^2}\left(\frac{1}{\rho_1}\frac{\partial Q_1}{\partial z} + \frac{1}{\rho_2}\frac{\partial Q_2}{\partial z}\right) = -\frac{\partial^2}{\partial x^2}\left(\frac{1}{\rho_1}\frac{\partial P_1^\circ}{\partial z} + \frac{1}{\rho_2}\frac{\partial P_2^\circ}{\partial z}\right), \quad (z = 0,\ x \to -0), \tag{3.6}$$

$$\frac{\partial^3}{\partial x^3}\left(\frac{1}{\rho_1}\frac{\partial Q_1}{\partial z} + \frac{1}{\rho_2}\frac{\partial Q_2}{\partial z}\right) = -\frac{\partial^3}{\partial x^3}\left(\frac{1}{\rho_1}\frac{\partial P_1^\circ}{\partial z} + \frac{1}{\rho_2}\frac{\partial P_2^\circ}{\partial z}\right), \quad (z = 0,\ x \to -0). \tag{3.7}$$

We seek the solution Q_1, Q_2 in the form

$$\begin{gathered}Q_1(x, z) = \int_{-\infty}^{+\infty} q_1(\lambda)\, e^{i\lambda x - s_1(\lambda) z}\, d\lambda,\\ Q_2(x, z) = \int_{-\infty}^{+\infty} q_2(\lambda)\, e^{i\lambda x + s_2(\lambda) z}\, d\lambda.\end{gathered} \tag{3.8}$$

The branches of the radicals are chosen as in § 2.

The boundary condition (3.5) reduces to the relation

$$\int_{-\infty}^{+\infty}\left[-\frac{s_1(\lambda)}{\rho_1} q_1(\lambda) - \frac{s_2(\lambda)}{\rho_2} q_2(\lambda)\right] e^{i\lambda x}\, d\lambda = 0, \tag{3.9}$$

which is valid on the entire axis, so that

$$-\frac{s_1(\lambda)}{\rho_1} q_1(\lambda) = \frac{s_2(\lambda)}{\rho_2} q_2(\lambda) \equiv \Psi(\lambda). \tag{3.10}$$

In order to ensure continuity of $Q_{1,2}(x, z)$ at $z=0$ we demand the following behavior of $\Psi(\lambda)$ at infinity:

$$\Psi(\lambda) \underset{\lambda\to\infty}{=} o(\lambda^{-\varepsilon}), \quad (\varepsilon > 0).$$

Using the boundary conditions (3.3) and (3.4), we obtain the following dual integral equation:

$$\begin{gathered}\int_{-\infty}^{+\infty} \left[\frac{\rho_1}{s_1(\lambda)} + \frac{\rho_2}{s_2(\lambda)}\right] \Psi(\lambda) e^{i\lambda x} d\lambda = 0 \quad (x>0), \\ \int_{-\infty}^{+\infty} \left\{ \left[\frac{\rho_1}{s_1(\lambda)} + \frac{\rho_2}{s_2(\lambda)} + 2m(\lambda)\right] \Psi(\lambda) + \frac{1}{\pi} \frac{m(\lambda) s_1(\lambda)}{\rho_2 s_1(\lambda) + \rho_1 s_2(\lambda)} e^{-i\lambda x_0 + s_2(\lambda) z_0} \right\} e^{i\lambda x} d\lambda = 0, \\ (x<0),\end{gathered} \tag{3.11}$$

where $m(\lambda)$ is obtained from (1.5) by the formal substitution of λ for $i\frac{\partial}{\partial x}$.

In order to satisfy the first equation of (3.11) we put

$$\left[\frac{\rho_1}{s_1(\lambda)} + \frac{\rho_2}{s_2(\lambda)}\right] \Psi(\lambda) = F^+(\lambda), \tag{3.12}$$

where $F^+(\lambda)$ is an analytic function in the upper half of the complex plane of λ. The second equation of (3.11) yields analogously

$$\left[\frac{\rho_1}{s_1(\lambda)} + \frac{\rho_2}{s_2(\lambda)} + 2m(\lambda)\right] \Psi(\lambda) + \frac{1}{\pi} \frac{m(\lambda) s_1(\lambda)}{\rho_2 s_1(\lambda) + \rho_1 s_2(\lambda)} e^{-i\lambda x_0 + s_2(\lambda) z_0} = F^-(\lambda), \tag{3.13}$$

where F^- is a function analytic in the lower half-plane. From (3.12) and (3.13) we arrive at the following Riemann boundary-value problem [8]: Find two functions $F^+(\lambda)$ and $F^-(\lambda)$ analytic in the upper and lower half-planes, from the following condition on the real axis:

$$\ell_2(\lambda) F^+(\lambda) = F^-(\lambda) + f_2(\lambda), \tag{3.14}$$

in which it is assumed that

$$\ell_2(\lambda) = 1 + 2 \frac{s_1(\lambda) \cdot s_2(\lambda) m(\lambda)}{\rho_1 s_2(\lambda) + \rho_2 s_1(\lambda)}, \tag{3.15}$$

$$f_2(\lambda) = \frac{1}{\pi} \frac{m(\lambda) \cdot s_1(\lambda)}{\rho_1 s_2(\lambda) + \rho_2 s_1(\lambda)} e^{-i\lambda x_0 + s_2 z_0}. \tag{3.16}$$

Moreover, the functions $F^+(\lambda)$ and $F^-(\lambda)$ must have the following order at infinity:

$$F^+(\lambda) = o(\lambda^{-1-\varepsilon}), \quad F^-(\lambda) = o(\lambda^{4-\varepsilon}), \quad (\varepsilon > 0). \tag{3.17}$$

For the solution of problem (3.14) we must first represent the function $\ell_2(\lambda)$ as a product of two functions, one of which $[\ell_2^+(\lambda)]$ is analytic in the upper half-plane, the other $[\ell_2^-(\lambda)]$ in the lower half-plane:

$$\ell_2(\lambda) = \ell_2^+(\lambda) \cdot \ell_2^-(\lambda). \tag{3.18}$$

The representation (3.18) (factorization) will be formulated in § 4. We shall also verify in that section that $\ell_2^{\pm}(\lambda)$ have the following order at infinity:

$$\ell_2^{\pm}(\lambda) = O(\lambda^{\frac{5}{2}}). \tag{3.19}$$

Taking (3.18) into account, we transform Eq. (3.14) as follows:

$$\ell_2^{+}(\lambda)\, F^{+}(\lambda) - \chi_2^{+}(\lambda) = \frac{F^{-}(\lambda)}{\ell_2^{-}(\lambda)} - \chi_2^{-}(\lambda), \tag{3.20}$$

in which it is assumed that

$$\chi_2(\lambda) = \frac{1}{2\pi i}\int_{-\infty}^{+\infty} \frac{1}{\pi \ell_2^{-}(\tau)} \cdot \frac{m(\tau)\cdot s_1(\tau)}{\rho_2 s_1(\tau) + \rho_1 s_2(\tau)}\, e^{-i\tau x_0 + s_2(\tau) z_0} \frac{d\tau}{\tau - \lambda} = \begin{cases} \chi_2^{+}(\lambda) & \text{for } \operatorname{Im}\lambda > 0, \\ \chi_2^{-}(\lambda) & \text{for } \operatorname{Im}\lambda < 0, \end{cases} \tag{3.21}$$

and the equations of Sokhotskii [8] are taken into account.

According to the Riemann theorem on analytic continuation through a contour the left- and right-hand sides of Eq. (3.20) give a unified function analytic in the entire plane of λ, and according to a theorem of Liouville [taking Eqs. (3.17), (3.19), and (3.20) into account] this function is a first-degree polynomial. We then have from (3.12)

$$\psi(\lambda) = \frac{s_1(\lambda)\; s_2(\lambda)}{\rho_1 s_2(\lambda) + \rho_2 s_1(\lambda)} \cdot \frac{\chi_2^{+}(\lambda) + a_0 + a_1 i\lambda}{\ell_2^{+}(\lambda)}, \tag{3.22}$$

where the constants a_0 and a_1 are arbitrary for the time being.

Equations (3.8), (3.10), and (3.22) give the diffraction fields:

$$\begin{aligned} Q_1(x,z) &= \int_{-\infty}^{+\infty} \frac{-\rho_1 s_2(\lambda)}{\rho_1 s_2(\lambda) + \rho_2 s_1(\lambda)} \cdot \frac{\chi_2^{+}(\lambda) + a_0 + a_1 i\lambda}{\ell_2^{+}(\lambda)}\, e^{i\lambda x - s_1(\lambda) z}\, d\lambda, \\ Q_2(x,z) &= \int_{-\infty}^{+\infty} \frac{\rho_2 s_1(\lambda)}{\rho_1 s_2(\lambda) + \rho_2 s_1(\lambda)} \cdot \frac{\chi_2^{+}(\lambda) + a_0 + a_1 i\lambda}{\ell_2^{+}(\lambda)} \times e^{i\lambda x + s_2(\lambda) z}\, d\lambda. \end{aligned} \tag{3.23}$$

Equations (3.1), (2.10), and (3.23) give the total field.

The constants a_0 and a_1 are determined from the boundary-contact conditions (3.6) and (3.7). However, the direct application of the boundary-contact operators to the field P_1, P_2 generates divergent integrals of expressions that exhibit algebraic growth at infinity. A method is presented in [7] for the regularization of such integrals. It is quickly seen that the boundary-contact operators can be applied to the field in the entire space of P_1°, P_2°, because with passage to the limit as $z=0$, $x \to -0$ in the integrands the exponentially decaying factors $e^{s_2(\lambda) z_0}$ are preserved. Consequently, the results of applying the boundary-contact operators to the diffraction field Q_1, Q_2 are subject to regularization.

We have

$$\frac{1}{\rho_1}\frac{\partial^2}{\partial x^2}\frac{\partial Q_1}{\partial z}(z=0,\, x\to -0) = \int_{-\infty}^{+\infty} \frac{(i\lambda)^2 s_1(\lambda) s_2(\lambda)}{\rho_1 s_2(\lambda) + \rho_2 s_1(\lambda)} \cdot \frac{\chi_2^{+}(\lambda) + a_0 + a_1 i\lambda}{\ell_2^{+}(\lambda)}\, e^{-i0\lambda}\, d\lambda, \tag{3.24}$$

$$\frac{1}{\rho_1}\frac{\partial^3}{\partial x^3}\frac{\partial Q_1}{\partial z}(z=0,\, x\to -0) = \int_{-\infty}^{+\infty} \frac{(i\lambda)^3 s_1(\lambda) s_2(\lambda)}{\rho_1 s_2(\lambda) + \rho_2 s_1(\lambda)} \cdot \frac{\chi_2^{+}(\lambda) + a_0 + a_1 i\lambda}{\ell_2^{+}(\lambda)}\, e^{-i0\lambda}\, d\lambda, \tag{3.25}$$

in which the following notation is used:

$$\int_{-\infty}^{+\infty} \varphi(\lambda)\, e^{-i0\lambda}\, d\lambda = \lim_{x \to -0} \int_{-\infty}^{+\infty} \varphi(\lambda)\, e^{i\lambda x}\, d\lambda. \tag{3.26}$$

The results of applying the boundary-contact operators to the field Q_2 are written analogously.

We denote

$$T_n(x) = \int_{-\infty}^{+\infty} \frac{(i\lambda)^n s_1(\lambda) s_2(\lambda)}{\rho_1 s_2(\lambda) + \rho_2 s_1(\lambda)} \cdot \frac{\chi_2^+(\lambda)}{\ell_2^+(\lambda)} e^{i\lambda x}\, d\lambda, \tag{3.27}$$

$$R_n(x) = \int_{-\infty}^{+\infty} \frac{(i\lambda)^n s_1(\lambda) s_2(\lambda)}{\rho_1 s_2(\lambda) + \rho_2 s_1(\lambda)} \cdot \frac{1}{\ell_2^+(\lambda)} e^{i\lambda x}\, d\lambda. \tag{3.28}$$

Then from the boundary-contact conditions (3.6) and (3.7) we deduce

$$T_2(-0) + a_0 R_2(-0) + a_1 R_3(-0) = \frac{1}{2\pi} \int_{-\infty}^{+\infty} \frac{s_1(\lambda) \cdot \lambda^2}{\rho_1 s_2(\lambda) + \rho_2 s_1(\lambda)} e^{-i\lambda x_0 + s_2(\lambda) z_0}\, d\lambda, \tag{3.29}$$

$$T_3(-0) + a_0 \cdot R_3(-0) + a_1 R_4(-0) = \frac{1}{2\pi} \int_{-\infty}^{+\infty} \frac{s_1(\lambda) \cdot \lambda^3}{\rho_1 s_2(\lambda) + \rho_2 s_1(\lambda)} e^{-i\lambda x_0 + s_2(\lambda) z_0}\, d\lambda. \tag{3.30}$$

We concern ourselves next with the regularization of the integrals (3.27) and (3.28). We have

$$T_n(x) = \int_{-\infty}^{+\infty} \frac{(i\lambda)^n \cdot s_1(\lambda) \cdot s_2(\lambda)}{\rho_1 s_2(\lambda) + \rho_2 s_1(\lambda)} \cdot \frac{[\chi_2^-(\lambda) + f_2(\lambda)] \cdot \ell_2^-(\lambda)}{\ell_2(\lambda)} e^{i\lambda x}\, d\lambda =$$

$$= \int_{-\infty}^{+\infty} \frac{(i\lambda)^n \cdot s_1(\lambda) \cdot s_2(\lambda)}{\rho_1 s_2(\lambda) + \rho_2 s_1(\lambda)} \cdot \frac{f_2(\lambda)}{\ell_2^+(\lambda)} e^{i\lambda x}\, d\lambda + \int_{-\infty}^{+\infty} \frac{(i\lambda)^n \cdot s_1(\lambda) \cdot s_2(\lambda)}{\rho_1 s_2(\lambda) + \rho_2 s_1(\lambda)} \cdot \frac{\chi_2^-(\lambda) \cdot \ell_2^-(\lambda)}{\ell_2(\lambda)} e^{i\lambda x}\, d\lambda. \tag{3.31}$$

The first integral (3.21) converges in the usual sense, because $f_2(\lambda)$ decays exponentially at infinity.

We deform the contour of integration in the second integral (3.31) into the lower half-plane until it encircles the branch cut (Fig. 2). We take account of the fact that the poles of the integrand were intercepted in deformation of the contour:

$$\int_{-\infty}^{+\infty} \frac{(i\lambda)^n s_1(\lambda) s_2(\lambda)}{\rho_1 s_2(\lambda) + \rho_2 s_1(\lambda)} \cdot \frac{\chi_2^-(\lambda) \cdot \ell_2^-(\lambda)}{\ell_2(\lambda)} e^{i\lambda x}\, d\lambda = -2\pi i \sum_{\operatorname{Im}\lambda < 0} \operatorname{Res} \frac{(i\lambda)^n s_1(\lambda) s_2(\lambda)}{\rho_1 s_2(\lambda) + \rho_2 s_1(\lambda)} \cdot \frac{\chi_2^-(\lambda) \cdot \ell_2^-(\lambda)}{\ell_2(\lambda)} e^{i\lambda x} +$$

$$+ \int_{L^+ \cup \Gamma^+} \frac{(i\lambda)^n s_1(\lambda) s_2(\lambda)}{\rho_1 s_2(\lambda) + \rho_2 s_1(\lambda)} \cdot \frac{\chi_2^-(\lambda) \cdot \ell_2^-(\lambda)}{\ell_2(\lambda)} e^{i\lambda x}\, d\lambda, \tag{3.32}$$

where $L^+ = L_1^+ \cup L_2^+$ and $\Gamma^+ = \Gamma_1^+ \cup \Gamma_2^+$ (Fig. 2). We reduce the integral over Γ^+ to an integral over Γ_1^+, and reduce the integral over L^+ to an integral over L_1^+. We obtain as a result

$$T_n(x) = \int_{-\infty}^{+\infty} \frac{(i\lambda)^n s_1(\lambda) s_2(\lambda)}{\rho_1 s_2(\lambda) + \rho_2 s_1(\lambda)} \cdot \frac{f_2(\lambda)}{\ell_2^+(\lambda)} e^{i\lambda x}\, d\lambda - 2\pi i \cdot \sum_{\operatorname{Im}\lambda < 0} \operatorname{Res} \frac{(i\lambda)^n s_1(\lambda) \cdot s_2(\lambda)}{[\rho_1 s_2(\lambda) + \rho_2 s_1(\lambda)]\, \ell_2^+(\lambda)} e^{i\lambda x} +$$

$$+2\cdot\int\limits_{L_1^+}\frac{\rho_2(i\lambda)^n\cdot s_1^2(\lambda)\cdot s_2(\lambda)\cdot\chi_2^-(\lambda)\cdot \ell_2^-(\lambda)}{\rho_2^2 s_1^2(\lambda)-[\rho_1 s_2(\lambda)+2s_1(\lambda)s_2(\lambda)m(\lambda)]^2}e^{i\lambda x}d\lambda+$$

$$+2\cdot\int\limits_{\Gamma_1^+}\frac{(i\lambda)^n\cdot s_1(\lambda)\cdot s_2(\lambda)\cdot\chi_2^-(\lambda)\cdot \ell_2^-(\lambda)\cdot[\rho_2 s_1(\lambda)+\rho_1 s_2(\lambda)]}{[\rho_1 s_2(\lambda)+\rho_2 s_1(\lambda)]^2-4s_1^2(\lambda)s_2^2(\lambda)\cdot m^2(\lambda)}e^{i\lambda x}d\lambda. \tag{3.33}$$

It is readily verified that the integrals over Γ_1^+ converge in the usual sense even without the factor $e^{i\lambda x}$ for $n\leq 6$, because the integrand has order $O(\lambda^{n-7.5})$ at infinity. We have thus effected regularization of the integrals $T_n(-0)$.

The integrals $R_n(-\infty)$ are regularized similarly. The final result is obtained from Eq. (3.32), in which it is required to make the formal substitution $f_2(\lambda)\equiv 0$, $\chi_2^-(\lambda)\equiv 1$. Ultimately the regulatization of $R_n(-0)$ is realized for $n\leq 5$.

The total field is now obtained from Eqs. (3.1), (2.10), (3.23), (3.29), and (3.30).

Next we consider, by analogy with the foregoing, the case of a perfectly rigid plate. The total field is obtained from Eqs. (3.1), (2.10), and (3.23) with the substitution therein of $a_0=a_1=0$ and the replacement of $\ell_2(\lambda)$ and $\chi_2(\lambda)$ by $\ell_1(\lambda)$ and $\chi_1(\lambda)$, where

$$\ell_1(\lambda)=\frac{\rho_1 s_1(\lambda)s_2(\lambda)}{\rho_1 s_2(\lambda)+\rho_2 s_1(\lambda)}=\ell_1^+(\lambda)\cdot\ell_1^-(\lambda),$$

$$\chi_1(\lambda)=\frac{1}{2\pi i}\int\limits_{-\infty}^{+\infty}\frac{1}{2\pi}\cdot\frac{\ell_1^+(\lambda)}{s_2(\tau)}e^{s_2(\tau)z_0-i\tau x_0}\frac{d\tau}{\tau-\lambda}=\begin{cases}\chi_1^+(\lambda)\text{ for }\operatorname{Im}\lambda>0,\\ \chi_1^-(\lambda)\text{ for }\operatorname{Im}\lambda<0,\end{cases} \tag{3.34}$$

the function $\ell_1^+(\lambda)$ is analytic in the upper half-plane, and $\ell_1^-(\lambda)$ is analytic in the lower half-plane. The representation (3.34) will be derived in § 4.

§ 4. Factorization of the Functions $\ell_{1,2}(\lambda)$

Here we derive the representations

$$\ell_{1,2}(\lambda)=\ell_{1,2}^+(\lambda)\ \ell_{1,2}^-(\lambda). \tag{4.1}$$

We use the apparatus of the Cauchy integrals [6, 7, 8]. The functions $\ell_{1,2}(\lambda)$ are determined on a four-sheeted Riemannian surface, on which they have four branch points $\lambda=\pm\kappa_1$, $\pm\kappa_2$. The branch cuts indicated in Fig. 2 separate the sheets of this surface. It is assumed in this connection that the cuts do not pass through the roots of the function $\ell_{1,2}(\lambda)$. We reduce the factorization of the functions $\ell_{1,2}(\lambda)$ to the factorization of a product of certain standard functions. We first consider $\ell_2(\lambda)$:

$$\ell_2(\lambda)=1+2m(\lambda)\frac{s_1(\lambda)\ s_2(\lambda)}{\rho_1 s_2(\lambda)+\rho_2 s_1(\lambda)}. \tag{4.2}$$

We introduce the function

$$\gamma(\lambda)\equiv\frac{s_1(\lambda)+s_2(\lambda)}{\sqrt{\kappa_1^2-\kappa_2^2}}. \tag{4.3}$$

Then, as we can readily verify, the following equations hold:

$$s_1(\lambda)=\frac{\sqrt{k_1^2-k_2^2}}{2}\left(\gamma-\frac{1}{\gamma}\right),$$

$$s_2(\lambda)=\frac{\sqrt{k_1^2-k_2^2}}{2}\left(\gamma+\frac{1}{\gamma}\right), \tag{4.4}$$

$$\lambda^2=\frac{1}{4}\left[k_1^2\left(\gamma+\frac{1}{\gamma}\right)^2-k_2^2\left(\gamma-\frac{1}{\gamma}\right)^2\right].$$

The function $l_2(\lambda)$ now assumes the form

$$l_2(\lambda)=\frac{1}{48\omega^2\gamma^5(\rho_1+\rho_2)}\cdot\frac{W(\gamma)}{\gamma^2+\frac{\rho_1-\rho_2}{\rho_1+\rho_2}}, \tag{4.5}$$

in which we have put

$$W(\gamma)=48\omega^2\gamma^5[\rho_1(\gamma^2+1)+\rho_2(\gamma^2-1)]+\sqrt{k_1^2-k_2^2}\,(\gamma^4-1)\times$$
$$\times\{48\omega^2h\rho_0\gamma^4-E_0h^3[(k_1^2-k_2^2)\gamma^4+2(k_1^2+k_2^2)\gamma^2+(k_1^2-k_2^2)]^2\}. \tag{4.6}$$

We denote the roots of the polynomial $W(\gamma)$ by γ_j $(j=1,2,\ldots,12)$. We have in this notation

$$l_2(\gamma)=-\frac{E_0h^3(k_1^2-k_2^2)^{5/2}}{48\omega^2\gamma^5(\rho_1+\rho_2)\cdot\left(\gamma^2+\frac{\rho_1-\rho_2}{\rho_1+\rho_2}\right)}\prod_{j=1}^{12}(\gamma-\gamma_j). \tag{4.7}$$

We introduce the function

$$E(\lambda,a)\equiv s_1(\lambda)+s_2(\lambda)-a\sqrt{k_1^2-k_2^2}. \tag{4.8}$$

We then have for $l_2(\lambda)$

$$l_2(\lambda)=-\frac{E_0h^3\prod_{j=1}^{12}E(\lambda,\gamma_j)}{48\omega^2(\rho_1+\rho_2)\,E\left(\lambda,\sqrt{\frac{\rho_2-\rho_1}{\rho_2+\rho_1}}\right)E\left(\lambda,-\sqrt{\frac{\rho_2-\rho_1}{\rho_2+\rho_1}}\right)[E(\lambda,0)]^5}. \tag{4.9}$$

The factorization of $l_2(\lambda)$ is thus reduced to the determination of the numbers γ_j and the factorization of $E(\lambda,a)$ for various a.

If the parameters of the media and plate and the oscillation frequency are given, the γ_j can be found either numerically or by the method of successive approximations in the form of a power series on some small parameter (usually the dimensionless plate thickness is adopted as the small parameter).

It is readily verified that for real wave numbers $k_{1,2}$ the function $l_2(\lambda)$ has only a pair of real zeros, which have an absolute value greater than $\max(k_1,k_2)$ and differ in sign. These roots are the wave numbers of surface modes. The other 11 pairs of roots of $l_2(\lambda)$ correspond to inhomogeneous waves.

The factorization of $E(\lambda,a)$ is realized by different techniques, depending on whether or not this function has zeros on the principal sheet, and this question is resolved, in turn, by the quantity a.

The function $E(\lambda,a)$ has two roots at the points

$$\lambda_a=\pm\frac{\sqrt{(a^2+k_1^2+k_2^2)^2-4k_1^2k_2^2}}{2a}, \tag{4.10}$$

where, by the evenness of $E(\lambda, a)$, both roots are located simultaneously on one of the four sheets.

Let us assume first that the pair of roots of $E(\lambda, a)$ is situated on the bottom sheet. Then the function

$$\varphi_1(\lambda, a) = 2\frac{E(\lambda, a)\, S_2(\lambda)}{\lambda^2 - \lambda_a^2} \tag{4.11}$$

does not have roots on the principal sheet, and

$$\lim_{\lambda\to\infty} \varphi_1(\lambda, a) = 1 . \tag{4.12}$$

Consequently, it is possible on the principal sheet to determine a single-valued function $\ln \varphi_1(\lambda, a)$, where

$$\lim_{\lambda\to\infty} \ln \varphi_1(\lambda, a) = 0 . \tag{4.13}$$

According to the Cauchy theorem

$$\ln \varphi_1(\lambda, a) = \frac{1}{2\pi i}\int_C \frac{\ln \varphi_1(\tau, a)}{\tau - \lambda}\, d\tau, \tag{4.14}$$

where C is any contour on the principal sheet that does not intercept the branch cut or encircle the point $\tau = \lambda$ in the positive direction. Extending this contour, we obtain

$$\frac{1}{2\pi i}\int_C \frac{\ln \varphi_1(\tau, a)}{\tau - \lambda}\, d\tau = \frac{1}{2\pi i}\int_{C^+} \frac{\ln \varphi_1(\tau, a)}{\tau - \lambda}\, d\tau + \frac{1}{2\pi i}\int_{C^-} \frac{\ln \varphi_1(\tau, a)}{\tau - \lambda}\, d\tau, \tag{4.15}$$

where the contours $C_\pm = \Gamma_\pm \cup L_\pm$. Finally, the factorization of $E(\lambda, a)$ is given by the equation

$$E^\pm(\lambda, a) = \frac{\lambda \mp \lambda_a}{\sqrt{2}\sqrt{\lambda \pm K_2}} \exp\left[\frac{1}{2\pi i}\int_{C_\pm} \frac{\ln \varphi_1(\tau, a)}{\tau - \lambda}\, d\tau\right], \tag{4.16}$$

in which λ_a denotes the roots of $E(\lambda, a)$, for which $\operatorname{Im} \lambda_a < 0$.

The factorization is analogous for the case in which $E(\lambda, a)$ does not have roots on the bottom sheet. We have

$$E(\lambda, a) = 2S_2(\lambda)\, \varphi_2(\lambda, a), \tag{4.17}$$

where the function $\varphi_2(\lambda, a)$ has the same properties on the principal sheet as $\varphi_1(\lambda, a)$ and is factorized in the same way.

Thus, we arrive at the factorization of $\ell_2(\lambda)$:

$$\ell_2(\lambda) = i\sqrt{\frac{E_0 h^3}{48\omega^2(\rho_1 + \rho_2)}}\; \frac{\prod_{j=1}^{12} E^\pm(\lambda, \gamma_j)}{E^\pm\left(\lambda, \sqrt{\frac{\rho_2 - \rho_1}{\rho_2 + \rho_1}}\right) E^\pm\left(\lambda, -\sqrt{\frac{\rho_2 - \rho_1}{\rho_2 + \rho_1}}\right) \left[E^\pm(\lambda, 0)\right]^5} . \tag{4.18}$$

It is seen at once that for all a

$$E^\pm(\lambda, a) \underset{\lambda\to\infty}{=} O(\lambda^{1/2}), \tag{4.19}$$

whence it follows that

$$\ell_2^{\pm}(\lambda) \underset{\lambda \to 0}{=} O(\lambda^{5/2}). \tag{4.20}$$

We obtain the representation for $\ell_1(\lambda)$ in analogous fashion. In this case

$$\ell_1(\lambda) = \frac{2\rho_1}{\sqrt{\kappa_1^2-\kappa_2^2}\,(\rho_1+\rho_2)} \, \frac{s_1(\lambda)\cdot s_2(\lambda)\,\gamma}{\gamma^2 + \frac{\rho_1-\rho_2}{\rho_1+\rho_2}}, \tag{4.21}$$

$$\ell_1^{\pm}(\lambda) = \sqrt{\frac{2\rho_1}{\rho_1+\rho_2}} \cdot \frac{\sqrt{\lambda \pm \kappa_1}\,\sqrt{\lambda \pm \kappa_2}\; E^{\pm}(\lambda, 0)}{E^{\pm}\left(\lambda, \sqrt{\frac{\rho_2-\rho_1}{\rho_2+\rho_1}}\right) E^{\pm}\left(\lambda, -\sqrt{\frac{\rho_2-\rho_1}{\rho_2+\rho_1}}\right)}. \tag{4.22}$$

LITERATURE CITED

1. Kouzov, D. P., On the low-frequency motions of a thin elastic layer separating two fluids, in: Wave Diffraction and Propagation Problems, No. 6, Leningrad Univ. (LGU) (1966).
2. Molotkov, L. A., Propagation of low-frequency oscillations in liquid half-spaces separated by a thin elastic layer, in: Problems in the Dynamical Theory of Seismic Wave Propagation, No. 5, Leningrad Univ. (LGU) (1961), pp. 281-302.
3. Ivakin, B. N., Transmitted head waves and other waves in the case of a solid layer in a liquid, Trudy Geofiz. Inst., No. 35 (1956).
4. Lyamshev, L. M. Sound Reflection by Thin Shells and Plates in a Liquid, Izd. AN SSSR (1955).
5. Papadopoulos, V. M., Diffraction of pulses, Proc. Roy. Soc., A252(1271):520-537 (1959).
6. Maue, A. W., Die Beugung elastischer Wellen an der Halbebene, Z. Angew. Math. Mech., Vol. 33 Nos. 1-2 (1953).
7. Kouzov, D. P., Diffraction of a cylindrical underwater sound wave by the junction of two semi-infinite plates, Prikl. Matem. i Mekh., Vol. 33 No. 2 (1969).
8. Gakhov, F. D., Boundary-Value Problems, GIFML (1963).

PROPAGATION OF LONGITUDINAL ELASTIC OSCILLATIONS IN THE QUANTIZED-THICKNESS WAVE-FILM MODE

T. S. Kravtsova

In the present article we solve the dynamical equations of elasticity theory, formulating solutions that are concentrated in the vicinity of a surface consisting of extremals of the Fermi integral

$$\int \frac{ds}{a(M)}, \tag{1}$$

where a is the longitudinal wave propagation velocity.

§1. Let the wave process be described by the dynamical elasticity-theoretic equations, which have the form (1)-(6) of [1] in an arbitrary curvilinear coordinate system. As in [1], we introduce a curvilinear coordinate system α, τ, ν, associated as follows with a surface S consisting of extremals of the integral (1):

$$\vec{X}(M) = \vec{X}(\alpha,\tau) + \nu\vec{n}(\alpha,\tau), \tag{1.1}$$

where $\vec{X} = \vec{X}(\alpha,\tau)$ is the parametric representation of S, $\alpha = \mathrm{const}$ is a ray, and the parameter τ varies along the ray. We interpret the rays as extremals of the integral (1); $\vec{n}(\alpha,\tau)$ is the unit normal to S at the point (α,τ), and ν is the distance along the normal from an arbitrary point M in space to S.

We assume that the displacement vector $\vec{u}$ has the form

$$\vec{u} = e^{-i\omega[t-\tau(M)]}\vec{\varphi}, \tag{1.2}$$

where ω is the frequency of the given wave process and is the large parameter of the problem, t is the time, and the components φ^j of the vector $\vec{\varphi}$ satisfy conditions 1) and 2) of [1]. The coefficients of the equations and the parameters of the medium are assumed to be sufficiently smooth.

§2. We seek the components of the displacement vector $\varphi^j(M)$ in the form of series in reciprocal powers of the large parameter:

$$\varphi^j(M) = \sum_{k=0}^{\infty} V_k^j(\alpha,\tau,\nu)\,\omega^{-k/2}, \tag{2.1}$$

where

$$\eta = \sqrt{\omega}\,\nu, \quad j = \alpha, \tau, \nu.$$

We substitute the expressions for $\varphi^j(M)$ from Eqs. (2.1) into the elasticity-theoretic equations, having first expanded the coefficients of the latter in power series in η. Equating coefficients of like powers of the large parameter, we find that the principal terms V_0^α and V_0^ν of the transverse components φ^α and φ^ν are equal to zero, while the remaining terms satisfy the relations

$$V_{s+2}^\alpha = 2\frac{(\vec{x}_\alpha, \vec{n}_\tau)}{|\vec{x}_\alpha|^2}\eta V_{s+1}^{'\tau} + f_{s+2}^\alpha(\alpha,\tau)$$
$$V_{s+2}^\nu = -ia^2\frac{\partial V_{s+1}^\tau}{\partial \eta} + f_{s+2}^\nu(\alpha,\tau) \tag{2.2}$$
$$(s=-1, 0, 1, \ldots),$$

i.e., are expressed in terms of elements of the principal, longitudinal component $\varphi^\tau(M)$ and their derivatives.

For the elements of $\varphi^\tau(M)$, taking Eqs. (2.1) into account, we deduce the equation

$$L V_{s+2}^\tau = f_{s+2}^\tau(\alpha,\tau) \qquad (s=-2,-1,0,1,\ldots), \tag{2.3}$$

in which the operator L has the form

$$L = \frac{\partial^2}{\partial\eta^2} + \frac{2i}{a^2}\frac{\partial}{\partial\tau} + \frac{i}{a^2}\frac{\partial}{\partial\tau}\ln\left[(\lambda+2\mu)a|\vec{x}_\alpha|\right] - \left[\frac{a_{\nu\nu}}{a^3}\bigg|_{\nu=0} + \frac{3(\vec{x}_\alpha, \vec{n}_\tau)^2}{|\vec{x}_\alpha|^2 a^4}\right], \tag{2.4}$$

and the right-hand side f_{s+2}^τ, like f_{s+2}^α and f_{s+2}^ν from Eqs. (2.2), are second-order linear differential operators, whose coefficients are polynomials in η. They are determined by equations of the type (1.8) in [1].

Relations (2.2)-(2.3) are recursive and make it possible to calculate the components φ^j of the displacement vector up to terms of order $\omega^{-N/2}$, where N is any integer.

§ 3. The principal term V_0^τ of the longitudinal component φ^τ (zeroth approximation) is determined from Eqs. (2.3) with $s=-2$:

$$L V_0^\tau = 0. \tag{3.1}$$

This equation is solved exactly as the zeroth-approximation equation (2.1) of [1] for the transverse case, and the principal term V_0^τ turns out to be

$$V_0^\tau = \frac{1}{a}\sqrt{\frac{\psi}{\lambda+2\mu}}\sqrt{\frac{a}{|\vec{x}_\alpha|}}\exp\left[-\frac{i(2n+1)}{2}\int a^2\psi^2 d\tau - \frac{\xi^2}{2}\left(1+\frac{i\psi_\tau}{a^2\psi^3}\right)\right]H_n(\xi), \quad (n=0,1,2,\ldots), \tag{3.2}$$

where $\psi = \psi(\alpha,\tau)$ is a positive real function comprising the solution of an ordinary differential equation of the form (2.3) in [1], wherein it is required to replace the transverse velocity $b(M) = \sqrt{\frac{\mu}{\rho}}$ by the longitudinal wave velocity $a(M) = \sqrt{\frac{\lambda+2\mu}{\rho}}$, λ and μ are parameters of the elastic medium, ρ is its density, $\xi = \psi(\alpha,\tau)\eta$ and $H_n(\xi)$ are nth-order Hermite polynomials.

We seek the next-higher approximations for the longitudinal component $V^\tau(M)$ in the form

$$V_{s+2,k}^\tau = \frac{1}{a}\sqrt{\frac{\psi}{\lambda+2\mu}}\sqrt{\frac{a}{|\vec{x}_\alpha|}}\exp\left[-\frac{i(2n+1)}{2}\int a^2\psi^2 d\tau + \frac{\xi^2}{2}\left(1+\frac{i\psi_\tau}{a^2\psi^3}\right)\right]\sum_{k\le n+s+4} a_{s+2,k}^\tau(\alpha,\tau)H_k(\xi), \tag{3.3}$$

$$K = \begin{cases} 1, 3, 5, \ldots & \text{for even } n; \\ 0, 2, 4, \ldots & \text{for odd } n. \end{cases} \tag{3.3}$$

If we substitute the expressions for $V^{\tau}_{s+2,K}$ from Eqs. (3.3) into Eqs. (2.3), we obtain the following differential equation for the functions $a^{\tau}_{s+2,K}$ (α,τ):

$$\frac{\partial a^{\tau}_{s+2,K}}{\partial \tau} = -\frac{i}{2} a^2 p^{\tau}_{s+2,K},$$

where $p^{\tau}_{s+2,K}(\alpha,\tau)$ are the known coefficients of the polynomials on the right-hand side of Eq. (2.3).

We have as a result, up to an arbitrary function of α,

$$a^{\tau}_{s+2,K}(\alpha,\tau) = -\frac{i}{2}\int a^2 p^{\tau}_{s+2,K}\, d\tau. \tag{3.4}$$

The higher approximations for the transverse components φ^{α} and φ^{ν} of the displacement vector are calculated from Eqs. (2.2).

§4. Consequently, the displacement vector may be asymptotically (as $\omega \to \infty$) represented as follows, up to principal terms:

$$\vec{u} = e^{-i\omega[t-\tau(M)]}\left(\omega^{-\frac{1}{2}} |\vec{X}_{\alpha}| V^{\alpha}_{1},\ |\vec{X}_{\tau}| V^{\tau}_{0},\ \omega^{-1/2} V^{\nu}_{1}\right), \tag{4.1}$$

where

$$V^{\tau}_{0} = \frac{\chi(\alpha)}{a}\sqrt{\frac{\psi}{\lambda+2\mu}}\sqrt{\frac{a}{|\vec{X}_{\alpha}|}}\exp\left[-\frac{i(2n+1)}{2}\int a^2\psi^2 d\tau - \frac{i\psi_{\tau}}{2a^2\psi^3}\zeta^2\right] e^{-\frac{\zeta^2}{2}} H_n(\zeta), \tag{4.2}$$

$$V^{\alpha}_{1} = 2\frac{(\vec{X}_{\alpha}, \vec{n}_{\tau})}{|\vec{X}_{\alpha}|^2}\nu V^{\tau}_{0}, \quad V^{\nu}_{1} = -ia^2\frac{\partial V^{\tau}_{0}}{\partial \nu},$$
$$n = 0, 1, 2, \ldots, \qquad \tau(M) = \int^{M}\frac{ds}{a(M)} \tag{4.3}$$

[$\chi(\alpha)$ is an arbitrary function of α], i.e., the main particle displacement of the medium occurs in the direction of propagation of the wave front, so that the resulting solutions are properly called longitudinal.

The concentrated character of the solution follows from the form of V^{τ}_{0}. Thus, the functions $e^{-\frac{\zeta^2}{2}} H_n(\zeta) = Z$ are solutions of the Weber equation

$$Z'' + (2n+1-\zeta^2)Z = 0.$$

For integer-valued n and $|\zeta| \leqslant \sqrt{2n+1}$ they oscillate, while for $|\zeta| > \sqrt{2n+1}$ they tend monotonically to zero. Thus the thickness of the layer spanned by the oscillations is quantized, i.e., it assumes only a discrete set of values of the form

$$|\zeta| = \sqrt{2n+1}, \quad n = 0, 1, 2, \ldots$$

or, taking account of the form of ζ,

$$|\nu| = \frac{\omega^{-\frac{1}{2}}}{\psi(\alpha,\tau)}\sqrt{2n+1}, \quad n = 0, 1, 2, \ldots. \tag{4.4}$$

Solutions of the type (4.1) are called longitudinal wave films, and the thickness of the layer spanned by the oscillations is called the film thickness.

It can be shown that the energy of waves of this type is propagated along rays in the first approximation. Thus, if the wave process is described by the dynamical elasticity-theoretic equations, the energy density has the form

$$\frac{\rho}{2}\vec{u}_t\vec{u}_t^* + \frac{1}{2}\sigma_{i'j'}(\vec{u})G^{i'i}G^{j'j}, \tag{4.5}$$

in which ε_{ij} are the components of the strain tensor, σ_{ij} are the components of the stress tensor (see [1]), and (G^{ij}) is the inverse of the matrix (G_{ij}) of the metric tensor in the given coordinate system:

$$\sigma_{ij} = \lambda \operatorname{div}\vec{u}\, G_{ij} + 2\mu\, \varepsilon_{ij}(\vec{u})$$

(the asterisk * denotes the complex conjugate).

We now consider the energy dE contained inside the small volume

$$\alpha_0 \leq \alpha \leq \alpha_0 + d\alpha, \quad t \leq \tau \leq t + d\tau,$$
$$-\nu_0 \leq \nu \leq \nu_0, \quad \nu_0 = O(1) \tag{4.6}$$

(t is the time).

We use Eqs. (4.1)-(4.3) to calculate the energy. The principal energy term in the moving small volume (4.6) has the form (as $\omega\to\infty$)

$$dE = \omega^2|\chi(\alpha)|^2\left\{\frac{\rho}{2}\frac{\psi}{\lambda+2\mu}\frac{a}{|\vec{x}_\alpha|}\int_{-\nu_0}^{\nu_0}\left[e^{-\frac{\varsigma^2}{2}}H_n(\varsigma)\right]^2 d\varsigma\,|\vec{x}_\tau|\,d\tau\,|\vec{x}_\alpha|\,d\alpha + \right.$$
$$\left. + \frac{\lambda+2\mu}{2a^2}\frac{\psi}{\lambda+2\mu}\frac{a}{|\vec{x}_\alpha|}\int_{-\nu_0}^{\nu_0}\left[e^{-\frac{\varsigma^2}{2}}H_n(\varsigma)\right]d\varsigma\,|\vec{x}_\tau|\,d\tau\,|\vec{x}_\alpha|\,d\alpha\right\}.$$

Making use of the fact that $\varsigma = \sqrt{\omega}\,\psi(\alpha,\tau)\nu$, $|\vec{x}_\tau| = a$, and replacing the limits of integration by $\pm\infty$, we arrive at the equation

$$dE = \omega^{3/2}|\chi(\alpha)|^2\sqrt{2\pi}\, n!\, d\tau\, d\alpha. \tag{4.7}$$

This expression depends only on the ray. The energy in the small moving volume (4.6) is therefore invariant as this volume moves along the ray-extremals of the integral $\int\frac{ds}{a(M)}$ at the wave velocity $a(M) = \sqrt{\frac{\lambda+2\mu}{\rho}}$.

LITERATURE CITED

1. Babich, V. M., and Kravtsova, T. S., Propagation of tranverse elastic oscillations in the quantized-thickness wave-film mode, this volume, pp. 11-19.
2. Babich, V. M., and Lazutkin, V. F., Eigenfunctions concentrated near a closed geodesic, Topics in Mathematical Physics, Vol. 2, Consultants Bureau, New York (1968), pp. 9-18.
3. Whittaker, E. T., and Watson, G. N., Modern Analysis, Cambridge University Press (1927).

EIGENFUNCTIONS OF A CERTAIN INTEGRAL EQUATION

N. V. Kuznetsov

§ 1. Introduction

In the present article we investigate the asymptotic behavior (as $a \to +\infty$) of the eigenfunctions of the integral equation

$$\mu \psi(x) = \int_0^a \sqrt{x\xi}\, J_\nu(x\xi)\, \psi(\xi)\, d\xi, \tag{1.1}$$

where J_ν is a Bessel function of order ν; the fixed ν is positive and greater than 1/2.

It follows at once from the Hilbert–Schmidt theory of integral equations with continuous symmetric kernels (see, e.g., [1], pp. 111-121) that there is a denumerable sequence of values of the parameter μ, viz., eigenvalues of the integral equation, for which (1.1) has a solution not identically equal to zero. We denote these solutions, which are the eigenfunctions of Eq. (1.1), by $\psi_n(x,a)$, $n = 0,1,2,\ldots$, numbering them in order of increasing number of zeros in the interval $(0, a)$. The fundamental result of our investigation comprises asymptotic equations for the functions $\psi_n(x,a)$, normalized by the condition

$$\int_0^a \psi_n^2(x,a)\,dx = 1 \tag{1.2}$$

[as we are aware, the functions $\psi_n(x,a)$ can always be made real for $0 \leq x \leq a$, since all eigenvalues of an integral equation with a real symmetric kernel are real]. The estimates of the remaining terms in these equations are uniform on x and n; the same equations therefore describe the behavior of the eigenfunctions $\psi_n(x,a)$ as $n \to \infty$.

It is feasible to obtain asymptotic estimates for the eigenfunctions of Eq. (1) by virtue of the fact that this equation has the same eigenfunctions as the boundary-value problem

$$\left(\frac{d}{dx}(x^2 - a^2)\frac{d}{dx} + a^2x^2 + \left(\nu^2 - \frac{1}{4}\right)\frac{a^2}{x^2}\right)\psi = \tilde{\lambda}\psi, \quad |\psi(0)|, |\psi(a)| < \infty. \tag{1.3}$$

Using the "standard-equation" principle, we obtain estimates, uniform on x and $\tilde{\lambda}$, for the solutions of the differential equation (1.3) for large a. This result enables us to obtain equations, uniform on order number, for the eigenvalues $\tilde{\lambda}_0, \tilde{\lambda}_1, \tilde{\lambda}_2, \ldots$ of problem (1.3) and asymptotic equations for the eigenvalues μ_0, μ_1, μ_2 of Eq. (1.1) for large a. We thereby obtain an over-all picture of the behavior of the eigenfunctions and eigenvalues of Eq. (1.1) for large a; it may be expostulated as follows in general terms.

For values of n small in comparison with a^2,

$$\mu_n \approx (-1)^n, \quad \tilde{\lambda}_n \approx (4n + 2\nu + 2)\,a^2. \tag{1.4}$$

In particular, for every fixed n,

$$\lim_{a\to\infty} \mu_n = (-1)^n, \quad \lim_{a\to\infty} \frac{\tilde{\lambda}_n}{a^2} = 4n+2\nu+2.$$

If now $n\to\infty$, then $\tilde{\lambda}_n \sim n^2$, and the eigenvalues μ_n rapidly decay: $|\mu_n| = \exp(-n\ln n + O(n))$. We note that only exponential decay can be guaranteed for the eigenvalues of a symmetric analytic kernel in the general case (as $e^{-\beta n}$ with fixed $\beta>0$; see [2], p. 238). The eigenfunctions $\Psi_n(x,a)$ for every fixed n (and uniformly on x in each finite interval) have the limit $\varphi_n(x)$ as $a\to\infty$ The limit function $\varphi_n(x)$ satisfies the differential equation

$$\frac{d^2\varphi_n}{dx^2} + \left(4n+2\nu+2-x^2-\frac{\nu^2-1/4}{x^2}\right)\varphi_n = 0, \tag{1.5}$$

which is obtained by formal passage to the limit as $a\to\infty$ from Eq. (1.3). In this case each of the functions $\varphi_n(x)$ coincides, up to the trivial factor $(-1)^n$, with its ν-order Hankel transform, i.e., satisfies the equation

$$(-1)^n \varphi_n(x) = \int_0^\infty \sqrt{x\xi}\, J_\nu(x\xi)\, \varphi_n(\xi)\, d\xi.$$

Besides, the functions $\varphi_n(x)$ form a complete orthonormal system in $\mathcal{L}_2(0,\infty)$; consequently, they play the same role in the theory of the Hankel transform as the Hermite functions do in the theory of the Fourier transform (see [3], pp. 102-111).

The integral equation (1.1) and the similar integral equation

$$\mu\Psi(x) = \frac{1}{\sqrt{2\pi}} \int_{-a}^{a} e^{ix\xi}\, \Psi(\xi)\, d\xi \tag{1.6}$$

[each eigenfunction of which coincides with one of the eigenfunctions of an equation of the form (2) with $\nu=\frac{1}{2}$ or $\nu=-\frac{1}{2}$] are natural adjuncts to the theory of open resonators and have therefore been studied by several authors (see [21, 23, 24]). The asymptotic properties of these functions have also been investigated (see [4], pp. 179-189, in which an extensive bibliography is given, and [5, 22, 25, 26]); we shall have little occasion, however, to use the results of these papers.

The author found it necessary to investigate the eigenfunctions of Eq. (1.1) in seeking to improve the estimate of the remainder term in the problem of the number of integral points in a circle.

§ 2. Eigenfunction Properties Deducible from the Integral Equation

2.1. Completeness and Regularity Properties. It is a well-known fact that the eigenfunctions of Eq. (1.1) form a complete orthonormal system in $\mathcal{L}_2(0,a)$, i.e., for $n\neq m$

$$\int_0^a \Psi_n(x,a)\, \Psi_m(x,a)\, dx = 0, \tag{2.1}$$

and the following equation holds for any function $f(x)$ with integrable modulus squared on $(0,a)$:

$$\lim_{N\to\infty} \int_0^a \left| f(x) - \sum_{n=0}^{N} \Psi_n(x,a) \int_0^a f(t)\, \Psi_n(t,a)\, dt \right|^2 dx = 0. \tag{2.2}$$

Also, since the function $z^{-\nu} J_\nu(z)$ is an entire function of z (see, e.g., [6], Chap. 17), the integral equation

$$\mu_n \Psi_n(x,a) = \int_0^a \sqrt{x\xi}\, J_\nu(x\xi)\, \Psi_n(\xi)\, d\xi \tag{2.3}$$

determines $x^{-1/2-\nu}\,\psi_n(x,a)$ as an entire function of the complex variable x (as is readily verified, of first order and finite type a).

2.2. Orthogonality on the Interval $[a,+\infty)$. We make use of the following fact from the theory of the Hankel transform of order ν (see [3], Chap. 8).

Theorem 1. Let $f(x)$ and $g(x)$ be functions from $\mathcal{L}_2(0,\infty)$, and let $F(x)$, and $G(x)$ be their νth-order Hankel transforms, i.e.,

$$F(x)=\int_0^\infty \sqrt{x\xi}\, J_\nu(x\xi) f(\xi)\,d\xi, \quad G(x)=\int_0^\infty \sqrt{x\xi}\, J_\nu(x\xi) g(\xi)\,d\xi, \tag{2.4}$$

where the integrals converge in the mean. Then the following Parseval equation holds:

$$\int_0^\infty F(x)\,G(x)\,dx = \int_0^\infty f(x)\,g(x)\,dx. \tag{2.5}$$

The following is an immediate consequence.

Theorem 2. Under the normalization (1.2)

$$\int_a^\infty \psi_n(x,a)\,\psi_m(x,a)\,dx = (1/\mu_n^2 - 1)\,\delta_{n,m}, \tag{2.6}$$

where $\delta_{n,m}=1$ for $n=m$ and $\delta_{n,m}=0$ for $n\neq m$.

Thus, Eq. (2.3) means that $\psi_n(x,a)$ is the Hankel transform of a function equal to $\mu_n^{-1}\psi_n(x,a)$ for $0\leq x\leq a$ and identically equal to zero for $x>a$. The Parseval equation therefore gives

$$\int_0^\infty \psi_n(x,a)\,\psi_m(x,a)\,dx = \frac{1}{\mu_n\mu_m}\int_0^a \psi_n(x,a)\,\psi_m(x,a)\,dx. \tag{2.7}$$

Equation (2.6) follows from this equation by virtue of (2.1). We obtain the following from (2.6) for $n=m$:

$$\frac{1}{\mu_n^2} = 1 + \int_a^\infty \psi_n^2(x,a)\,dx > 1. \tag{2.8}$$

Consequently, all the eigenvalues of Eq. (2.3) have a modulus smaller than unity for any positive a.

2.3. Integral Equation on the Interval $[a,\infty)$. We shall prove the following.

Theorem 3. Let $\psi_n(x)$ be the nth eigenfunction of Eq. (1.1), and let μ_n be the corresponding eigenvalue. Then

$$\int_a^\infty \sqrt{x\xi}\, J_\nu(x\xi)\,\psi_n(\xi,a)\,d\xi = \begin{cases} (1/\mu_n - \mu_n)\,\psi_n(x,a), & x<a; \\ (1/2\mu_n - \mu_n)\,\psi_n(a,a), & x=a; \\ -\mu_n\,\psi_n(x,a), & x>a. \end{cases} \tag{2.9}$$

For the proof we first write

$$\int_0^b \sqrt{x\xi}\, J_\nu(x\xi)\psi_n(\xi)\,d\xi = \frac{1}{\mu_n}\int_0^a \psi_n(t,a)\Big(\int_0^b \sqrt{x\xi}\,\sqrt{\xi t}\, J_\nu(x\xi) J_\nu(\xi t)\,d\xi\Big)dt \equiv \frac{1}{\mu_n}\int_0^a \sqrt{xt}\,\psi_n(t,a)\,H(x,t;b)\,dt, \tag{2.10}$$

in which (see [6], Example 18 on p. 224 and the recursion formulas on p. 194)

$$H(x,t;b)=\frac{b}{x^2-t^2}\left(xJ_{\nu+1}(bx)J_\nu(bt)-tJ_{\nu+1}(bt)J_\nu(bx)\right). \tag{2.11}$$

Moreover, as $b\to\infty$ the function $H(x,t;b)$ behaves as a delta function of the argument $x-t$. The latter proposition is stated more precisely in the following theorem.

Theorem 4. Let the function $f(t)$ have bounded variation at every point of the interval $[0,a]$. Then

$$\lim_{b\to\infty}\int_0^a\sqrt{xt}\,f(t)H(x,t;b)\,dt=\begin{cases}\frac{1}{2}f(x+0)+\frac{1}{2}f(x-0), & 0<x<a;\\ \frac{1}{2}f(a-0), & x=a;\\ 0, & x>a.\end{cases} \tag{2.12}$$

This is the Hankel theorem; its proof may be found, for example, in [3], pp. 331-314. It implies that the limit of the left-hand side of Eq. (2.10) as $b\to\infty$ exists and is equal to zero for $x>a$, to $\frac{1}{2\mu_n}\psi_n(a,a)$ for $x=a$, and to $\frac{1}{\mu_n}\psi_n(x,a)$ for $0<x<a$. Now, subtracting Eq. (2.3) from both sides of (2.10), we arrive at the statement of Theorem 4.

2.4. Differential Equation Satisfied by the Functions $\psi_n(x,a)$

Theorem 5. For every n, $n=0,1,2,\ldots$, there exists a positive number $\tilde\lambda_n$ such that

$$(x^2-a^2)\frac{d^2\psi_n}{dx^2}+2x\frac{d\psi_n}{dx}+a^2\left(x^2+\frac{\nu^2-1/4}{x^2}\right)\psi_n=\tilde\lambda_n\psi_n. \tag{2.13}$$

The proof of this theorem is given in [22]. We note that the number $\tilde\lambda_n$ is necessarily positive. Thus, it follows from (2.13) that

$$\tilde\lambda_n=\tilde\lambda_n\int_0^a\psi_n^2(x,a)\,dx=\int_0^a\psi_n(x,a)\left(\frac{d}{dx}(x^2-a^2)\frac{d\psi_n}{dx}+\left(a^2x^2+(\nu^2-1/4)\frac{a^2}{x^2}\right)\psi_n\right)dx=$$
$$=\int_0^a\left((a^2-x^2)\left(\frac{d\psi_n}{dx}\right)^2+\left(a^2x^2+\left(\nu^2-\frac{1}{4}\right)\frac{a^2}{x^2}\right)\psi_n^2\right)dx. \tag{2.14}$$

The right-hand side is clearly positive; moreover, inasmuch as $x^2+\frac{\nu^2-1/4}{x^2}\geqslant\sqrt{4\nu^2-1}$ (we recall that $\nu>\frac{1}{2}$), the following inequality is valid:

$$\tilde\lambda_n\geqslant a^2\sqrt{4\nu^2-1}. \tag{2.15}$$

2.5. Boundary-Value Problem (1.3).

We now show that the converse proposition is true, namely that every eigenfunction of the boundary-value problem (1.3) is an eigenfunction of the integral equation (1.1).

First of all, the boundary-value problem (1.3) has a discrete spectrum. Thus, the points $x=0$ and $x=a$ are regular singular points of Eq. (1.3), and the characteristic equation corresponding to the point $x=0$ has roots $\nu+\frac{1}{2}$ and $-\nu+\frac{1}{2}$, whereas zero is a multiple root of the characteristic equation corresponding to $x=a$. Consequently (see, e.g., [7], Chap. 4, §4 or §8), there exist solutions $W_\kappa(x,\tilde\lambda)$, $\kappa=1, 2, 3, 4$, such that

$$W_1(x,\tilde\lambda)=x^{\nu+1/2}(1+c_1x+\cdots),\quad W_2(x,\tilde\lambda)=x^{-\nu+1/2}(1+\tilde c_1x+\cdots), \tag{2.16}$$

where the power series converge in some neighborhood of the point $x=0$ and

$$W_3(x,\tilde{\lambda})=1+b_1(x-a)+\cdots,\quad W_4(x,\tilde{\lambda})=W_3(x,\tilde{\lambda})\ln(x-a)+\tilde{b}_0+\tilde{b}_1(x-a)+\cdots, \tag{2.17}$$

where the power series converge for sufficiently small values of $|x-a|$. We know that for every fixed $x\neq 0,a$ the functions $W_k(x,\tilde{\lambda})$ thus normalized are entire functions of $\tilde{\lambda}$. Moreover, the solutions W_1, W_3, and W_4 are dependent, and in the equation

$$W_1(x,\tilde{\lambda})=\ell_1(\tilde{\lambda})W_3(x,\tilde{\lambda})+\ell_2(\tilde{\lambda})W_4(x,\tilde{\lambda}) \tag{2.18}$$

the coefficients ℓ_1 and ℓ_2 are analytic functions of $\tilde{\lambda}$. It follows from (2.16)-(2.18) that (for $\nu>\frac{1}{2}$) Eq. (1.3) has a finite solution for $x\to 0$ and $x\to a$ only if $\tilde{\lambda}$ is a zero of the analytic function $\ell_2(\tilde{\lambda})$. Therefore, the eigenfunctions of problem (1.3) form a discrete set. In this case, as the Sturm comparison theorems at once imply (see, e.g., [8], pp. 142-150), the nth [in order of increasing number of zeros on the interval $(0,a)$] eigenfunction has exactly n zeros, $n=0,1,2,\ldots$, in $(0,a)$, and the corresponding eigenvalues form a strictly increasing sequence: $\tilde{\lambda}_0<\tilde{\lambda}_1<\cdots$

Let $\psi_n(x)$ be an eigenfunction of the boundary-value problem (1.3); we show that the function

$$\tilde{\psi}_n=\int_0^a\sqrt{x\xi}\,J_\nu(x\xi)\,\psi_n(\xi)\,d\xi \tag{2.19}$$

is also an eigenfunction of this problem, corresponding to the same eigenvalue $\tilde{\lambda}_n$. Thus, let

$$L_x=(x^2-a^2)\frac{d^2}{dx^2}+2x\frac{d}{dx}+a^2\left(x^2+\frac{\nu^2-1/4}{x^2}\right).$$

We can readily verify that

$$L_x\tilde{\psi}_n=(\xi^2-a^2)\left(\psi_n(\xi)\frac{d}{d\xi}\sqrt{x\xi}\,J_\nu(x\xi)-\sqrt{x\xi}\,J_\nu(x\xi)\frac{d}{d\xi}\psi_n(\xi)\right)\Big|_{\xi=0}^{\xi=a}+\int_0^a\sqrt{x\xi}\,J_\nu(x\xi)\,L_\xi\psi_n\,d\xi. \tag{2.20}$$

But when $\xi\to 0$ we have $\psi_n=o(\xi^{\nu+1/2})$, $\psi_n'=o(\xi^{\nu-1/2})$, and for $\xi=a$ these functions are regular. The terms outside the integral in (2.20) therefore vanish. Consequently, $\tilde{\psi}_n$ differs from ψ_n only by a constant factor, so that $\psi_n(x)$ is an eigenfunction of Eq. (1.1)

2.6. Differential Equation for the Eigenvalues

Theorem 6. *Every eigenvalue μ_n of the integral equation (1.1) is a differentiable function of a, and*

$$\frac{d\mu_n}{da}=\mu_n(a)\psi_n^2(a,a). \tag{2.21}$$

This equation is entirely analogous to the equation obtained by Fuchs in [5] for the eigenvalues of the integral equation (1.6). For the proof we multiply the equation

$$\mu_n(b)\psi_n(x,b)=\int_0^b\sqrt{x\xi}\,J_\nu(x\xi)\,\psi_n(\xi,b)\,d\xi \tag{2.22}$$

by $\psi_n(x,a)$ and multiply Eq. (2.3) by $\psi_n(x,b)$, subtract term by term, and integrate the result over x from 0 to a. We obtain

$$(\mu_n(b)-\mu_n(a))\int_0^a\psi_n(x,a)\psi_n(x,b)\,dx=\int_0^a\psi_n(x,a)\left(\int_a^b\sqrt{x\xi}\,J_\nu(x\xi)\psi_n(\xi,b)\,d\xi\right)dx. \tag{2.23}$$

We divide this equation by $b-a$; when $b\to a$ the limit of the right-hand side exists and is equal to

$$\int_0^a \Psi_n(x,a)\sqrt{ax}\,J_\nu(ax)\,\Psi_n(a,a)\,dx = \mu_n(a)\,\Psi_n^2(a,a). \tag{2.24}$$

Moreover,

$$\lim_{b\to a}\int_0^a \Psi_n(x,b)\Psi_n(x,a)\,dx = \int_0^a \Psi_n^2(x,a)\,dx = 1. \tag{2.25}$$

The derivative $\frac{d\mu_n(a)}{da}$ therefore exists, and Eq. (2.21) is valid.

It follows from (2.21) that the positive eigenvalues increase monotonically with increasing a, while the negative eigenvalues decrease monotonically. Since $\mu_n^2(a)<1$ for any a, $\lim_{a\to\infty}\mu_n$ exists for every fixed n; we shall demonstrate later that this limit is equal to $(-1)^n$

§ 3. Estimates of the Solutions of Certain Large-Parameter Differential Equations

3.1. Standard-Equation Method. We consider the second-order differential equation

$$\frac{d^2\psi}{dx^2}+q(x,a)\,\psi=0, \qquad x_1\le x\le x_2, \tag{3.1}$$

in which a is a large positive parameter and the function $q(x,a)$ is assumed to be real and sufficiently smooth for $x\in(x_1,x_2)$. One method for the asymptotic integration of Eq. (3.1) is as follows. We introduce in (3.1) the new independent variable $\xi=\xi(x)$, $\frac{d\xi}{dx}>0$, and the new unknown function $W(\xi)$ by way of the equation

$$\psi(x)=\left(\frac{d\xi}{dx}\right)^{-1/2}W(\xi(x)). \tag{3.2}$$

A direct calculation shows that $W(\xi)$ satisfies the equation

$$\frac{d^2W}{d\xi^2}+\left(\frac{dx}{d\xi}\right)^2\left(q-\frac{1}{2}\{\xi,x\}\right)W=0, \quad \xi(x_1)\le\xi\le\xi(x_2), \tag{3.3}$$

where $\{\xi,x\}$ denotes the Schwarzian of ξ on x:

$$\{\xi,x\}=\xi'''/\xi'-\frac{3}{2}(\xi''/\xi')^2, \qquad '=\frac{d}{dx}. \tag{3.4}$$

Let us assume that the function $\xi(x)$ can be chosen so that the coefficient in front of W in Eq. (3.3) is represented in the form $p(\xi,a)-f(\xi,a)$, where, first, $|f|$ is "small" in comparison with $|p|$ and, second, the solutions of the transformed equation with $f=0$ are known functions. In this case Eq. (3.3) reduces to an integral equation of the type

$$W(\xi)=c_1w_1(\xi)+c_2w_2(\xi)+\int_{\xi_0}^{\xi}\frac{w_1(\xi)w_2(t)-w_2(\xi)w_1(t)}{\mathfrak{W}[w_1,w_2]}f(t)W(t)\,dt, \tag{3.5}$$

where c_1 and c_2 are constants, w_1 and w_2 are two linearly independent solutions of the equation with $f=0$, and $\mathfrak{W}[w_1,w_2]$ denotes the Wronskian of the functions w_1 and w_2:

$$\mathfrak{W}[w_1,w_2]=w_1\frac{dw_2}{d\xi}-w_2\frac{dw_1}{d\xi}. \tag{3.6}$$

The usual resulting integral equation of the Volterra type can be solved by the method of successive approximations, even when the function $q(x,a)$ in (3.1) has singularities or the interval (x_1,x_2) is not bounded. Even the first term of the iterated series yields, for large values of the parameter, a rather good approximation for the solution, and this approximation is uniform on x in the interval (x_1,x_2).

The method described above for the asymptotic integration of a second-order equation is sometimes known as the "standard-equation" method. The method was first used by Liouville [10] as long ago as 1837 in his classical study of the Sturm – Liouville problem.* Liouville investigated an equation of the type (3.1) with $q = a^2 q_0(x) + q_1(x)$ where a is large, $q_1(x)$ is continuous, and the function $q_0(x)$ is twice-continuously differentiable and positive definite in a finite interval $[x_1, x_2]$. Setting $\xi(x) = \int^x \sqrt{q_0(t)}\, dt$, he reduced the problem of estimating the solutions of Eq. (3.1) to the simple problem of analyzing the behavior of the solutions of the equation

$$\frac{d^2 W}{d\xi^2} + a^2 W = f(\xi) W, \tag{3.7}$$

in which $f(\xi)$ is continuous for $\xi_1 \leq \xi \leq \xi_2$. Liouville's asymptotic equations for the solutions of this equation were improved by Blumenthal [12] and later by Olver [13], who deduced the following inequality.

Theorem 7. Let the function $|f(\xi)|$ in Eq. (3.7) be integrable in the interval (ξ_1, ξ_2). Then there are solutions W_1 and $W_2 = W_1^*$ such that for $\xi_1 < \xi < \xi_2$

$$|W_1 - e^{ia\xi}| \leq \exp\left(\frac{2}{a} \left| \int_c^{\xi} |f(t)|\, dt\right) - 1, \tag{3.8}$$

where c is an arbitrary point on (ξ_1, ξ_2).

Analogous inequalities can be obtained for more complex situations in which the function $q(x, a)$ in Eq. (3.1) has zeros in the interval (x_1, x_2) or either end of this interval is a singular point of the equation. The following is useful in all these situations.

Theorem 8. Let $w_1(\xi)$ and $w_2(\xi)$ be linearly independent solutions of the equation

$$\frac{d^2 w}{d\xi^2} + p(\xi) w = 0, \quad \xi_1 < \xi < \xi_2, \tag{3.9}$$

and let there exist functions $\varphi(\xi)$ and $\tilde{\varphi}(\xi)$ such that for $\xi \in (\xi_1, \xi_2)$

$$|w_1(\xi)| \leq \varphi(\xi), \qquad |w_2(\xi)| \leq \tilde{\varphi}(\xi). \tag{3.10}$$

Let there exist in the interval (ξ_1, ξ_2) a point c (possibly coinciding with ξ_1 or ξ_2) such that $\varphi(\xi)/\tilde{\varphi}(\xi)$ is nonincreasing in the interval (ξ_1, c) and is nondecreasing in the interval (c, ξ_2). Then for any function $f(\xi)$ for which the integrals

$$F_1(\xi) = \int_c^{\xi} \varphi(t)\tilde{\varphi}(t)|f(t)|\, dt, \quad F_2(\xi) = F_1(\xi_2) - F_1(\xi), \quad F_3(\xi) = F_1(\xi) - F_1(\xi_1)$$

are meaningful there exist solutions W_1, W_2, and W_2 of the equation

$$\frac{d^2 W}{d\xi^2} + p(\xi) W = f(\xi) W \tag{3.11}$$

such that

$$|W_1(\xi) - w_1(\xi)| \leq \varphi(\xi)\left(e^{2\alpha |F_1(\xi)|} - 1\right), \quad \xi_1 < \xi < \xi_2; \tag{3.12}$$

$$|W_2(\xi) - w_2(\xi)| \leq \tilde{\varphi}(\xi)\left(e^{2\alpha F_2(\xi)} - 1\right), \quad c \leq \xi < \xi_2; \tag{3.13}$$

*The transformation (3.2) was used simultaneously (and independently) by Green [11].

$$|W_3(\xi) - w_2(\xi)| \leq \tilde{\varphi}(\xi)\left(e^{2\alpha F_3(\xi)} - 1\right), \quad \xi_1 < \xi \leq c, \tag{3.14}$$

where $\alpha = |w_1 w_2' - w_2 w_1'|$.

We shall prove this theorem in subsection 3.2; for the time being we note the following. The interval (ξ_1, ξ_2) can also be infinite, and the point c can be a singular point of Eq. (3.9). Consequently, an integral equation of the type (3.5) obtained through the formal application of the method of variation of arbitrary constants is, generally speaking, singular. There are well-known inequalities analogous to (3.12)-(3.14) for the solutions of singular Volterra equations of this type (see, e.g., Erdélyi's paper [14], pp. 8-12); however, Theorem 8 is better suited to the estimation of the solutions of the differential equations.

3.2. Inequalities for the Solutions of the Second-Order Equation. We now prove inequality (3.12) of Theorem 8; the other inequalities are proved similarly. For this objective we consider the integral equation (3.5), in which we put $c_1 = 1$, $c_2 = 0$, and $\xi_0 = c$. Denoting the kernel of this equation by $A(x,t)$, we let

$$W^{(n)}(\xi) = \int_c^{\xi} A(\xi; t) W^{(n-1)}(t)\,dt, \quad W^{(0)}(\xi) = w_1(\xi). \tag{3.15}$$

The Wronskian of the solutions w_1 and w_2 is independent of ξ; denoting its modulus by $1/\alpha$, we have by virtue of (3.10)

$$|A(\xi, t)| \leq \alpha\varphi(\xi)\tilde{\varphi}(t)|f(t)|\left(1 + \frac{\tilde{\varphi}(\xi)}{\varphi(\xi)}\frac{\varphi(t)}{\tilde{\varphi}(t)}\right).$$

According to the premise of the theorem the function $\varphi(\xi)/\tilde{\varphi}(\xi)$ is nondecreasing for $\xi \in (c, \xi_2)$ and nonincreasing for $\xi \in (\xi_1, c)$, therefore, for $\xi \leq t \leq c$ and for $c \leq t \leq \xi$ the inequality $\varphi(t)\tilde{\varphi}(\xi) \leq \varphi(\xi)\tilde{\varphi}(t)$ is valid. Consequently, in Eqs. (3.15)

$$|A(\xi, t)| \leq 2\alpha\varphi(\xi)\tilde{\varphi}(t)|f(t)|. \tag{3.16}$$

The following inequality therefore holds for the function $W^{(1)}(\xi)$:

$$|W^{(1)}(\xi)| \leq 2\alpha\varphi(\xi)\left|\int_c^{\xi}\varphi(t)\tilde{\varphi}(t)|f(t)|dt\right| \leq 2\alpha\frac{|F_1(\xi)|}{1!}\varphi(\xi). \tag{3.17}$$

Also, if the inequality

$$|W^{(n)}(\xi)| \leq (2\alpha)^n\frac{|F_1(\xi)|^n}{n!}\varphi(\xi) \tag{3.18}$$

holds for some $n \geq 1$, we then obtain for $W^{(n+1)}$ by virtue of (3.16)

$$|W^{(n+1)}(\xi)| \leq (2\alpha)^{n+1}\varphi(\xi)\left|\int_c^{\xi}\varphi(t)\tilde{\varphi}(t)|f(t)|\frac{|F_1(t)|^n}{n!}dt\right|. \tag{3.19}$$

But $F_1(\xi)$ is sign-preserving on each of the intervals (ξ_1, c) and (c, ξ_2) and is equal to zero at $\xi = c$. It follows from (3.19), therefore, that

$$|W^{(n+1)}(\xi)| \leq (2\alpha)^{n+1}\frac{|F_1(\xi)|^{n+1}}{(n+1)!}\varphi(\xi). \tag{3.20}$$

Consequently, inequality (3.18) holds for all $n \geq 0$. Therefore, the series

$$W(\xi) = \sum_{n=0}^{\infty} W^{(n)}(\xi) \tag{3.21}$$

is majorized by the series for $(2\alpha|F_1(\xi)|)$ and is thus convergent absolutely and uniformly on ξ in every finite subinterval of the interval (ξ_1, ξ_2). Term-by-term integration of the series (3.21) is therefore admissible; the result is

$$\begin{aligned}\int_c^\xi A(\xi,t)W(t)dt &= \sum_{n=0}^{\infty}\int_c^\xi A(\xi,t)W^{(n)}(t)dt \\ &= \sum_{n=1}^{\infty} W^{(n)}(\xi) \\ &= W(\xi)-W^{(0)}(\xi).\end{aligned} \tag{3.22}$$

Consequently, the sum of the series (3.21) satisfies the integral equation (3.5) with $c_1=1$, $c_2=0$, and $\xi_0=c$, and twofold differentiation of Eq. (3.5) shows that its solution also satisfies Eq. (3.11). Finally, the resulting solution satisfies the inequality

$$|W(\xi)-w_1(\xi)| \leq \sum_{n=1}^{\infty}|W^{(n)}(\xi)| \leq \varphi(\xi)\sum_{n=1}^{\infty}\frac{(2\alpha)^n|F_1(\xi)|}{n!},$$

which coincides with (3.12).

<u>3.3. Example (Turning-Point Equation).</u> As an example illustrating Theorem 8 we consider the equation

$$\frac{d^2W}{d\xi^2} - a^4\xi W = f(\xi)W, \quad \xi_1<\xi<\xi_2 \quad (-\infty\leq\xi_1<\xi_2\leq\infty). \tag{3.23}$$

The problem of deducing asymptotic equations for the solutions of Eq. (3.1) reduces to the estimation of the solutions of the above equation when the coefficient in front of the large parameter in $q(x,a)$ has a simple zero. The Liouville–Green transformation* was first used in this problem by Langer [15]; its asymptotic representations were later improved by Olver [16] (see also [17], pp. 27-31, in which the same inequalities are obtained).

The solutions of the unperturbed equation (3.23) are expressed in terms of the Airy function $Ai(z)$ and the associated Airy function $Bi(z)$ (see [18] or [19]). These functions satisfy the equation

$$\frac{d^2w}{dz^2} = zw \tag{3.24}$$

and are related to the Bessel functions of order $\pm 1/3$ by the relations

$$3Ai(z) = \sqrt{z}\left(I_{-1/3}\left(\tfrac{2}{3}z^{3/2}\right) - I_{1/3}\left(\tfrac{2}{3}z^{3/2}\right)\right) \tag{3.25}$$

$$Bi(z) = e^{i\pi/6}Ai\left(ze^{\frac{2\pi i}{3}}\right) + e^{-i\pi/6}Ai\left(ze^{-\frac{2\pi i}{3}}\right). \tag{3.26}$$

The functions $Ai(z)$ and $Bi(z)$ are entire functions of z and are real when z is real. The behavior of $Ai(z)$ as $|z|\to\infty$ is described by the following equations (branches that are positive on the positive semiaxis are chosen for the fractional powers)

$$Ai(z) = \begin{cases} \frac{1}{2\sqrt{\pi}} z^{-1/4}\exp\left(-\frac{2}{3}z^{3/2}\right)\left(1+O\left(\frac{1}{z^{3/2}}\right)\right), & |\arg z|<\pi, \quad (3.27) \\ \frac{1}{\sqrt{\pi}}(-z)^{-1/4}\cos\left(\frac{2}{3}(-z)^{3/2} - \frac{\pi}{4} + O\left(\frac{1}{z^{3/2}}\right)\right), & |\arg(-z)|\leq\frac{\pi}{3}. \quad (3.28)\end{cases}$$

*The name customarily given to the transformation from Eq. (3.1) to (3.3).

Asymptotic formulas for $Bi(z)$ are obtained by the substitution of the above equations into (3.26). These equations imply, in particular, that for all real x, $-\infty < x < \infty$,

$$|Ai(x)| \le k \frac{|\exp(-\frac{2}{3}x^{3/2})|}{1+|x|^{1/4}}, \quad |Bi(x)| \le k \frac{|\exp(\frac{2}{3}x^{3/2})|}{1+|x|^{1/4}}, \tag{3.29}$$

where κ is an absolute constant. Also solutions of Eq. (3.23) with $f \equiv 0$ are the functions $Ai(a^{4/3}\xi)$ and $Bi(a^{4/3}\xi)$, and the direct application of Theorem 8 produces the following result.

Theorem 9. In Eq. (3.23) let $a>0$, and let the function $(1+\sqrt{|\xi|})^{-1}|f(\xi)|$ be integrable in the interval (ξ_1, ξ_2). Then there exist solutions u_1 and u_2 of Eq. (3.23) such that for $\xi_1 < \xi < \xi_2$

$$(1+a^{1/3}|\xi|^{1/4})|u_1 - Ai(a^{4/3}\xi)| \le \kappa |\exp(-\tfrac{2}{3}a^2\xi^{3/2})| (e^{F_1(\xi)}-1), \tag{3.30}$$

$$(1+a^{1/3}|\xi|^{1/4})|u_2 - Bi(a^{4/3}\xi)| \le \kappa |\exp(\tfrac{2}{3}a^2\xi^{3/2})| (e^{F_2(\xi)}-1), \tag{3.31}$$

where

$$F_1(\xi) = \frac{4\pi\kappa^2}{a^{4/3}} \int_{\xi}^{\xi_2} \frac{|f(t)|\,dt}{(1+a^{1/3}|t|^{1/4})^2}, \quad F_2(\xi) = F_1(\xi_1) - F_1(\xi). \tag{3.32}$$

These inequalities are also given in the above-mentioned paper of Olver [16].

3.4. Approximation by Bessel Functions. In the event an equation of the form (3.1) has a regular singular point [and there are no other singularities or zeros of the function $q(x, a)$ in the vicinity of that point] the following equations may be chosen as the standard:

$$W'' + (a^4 - \frac{\nu^2 - 1/4}{\xi^2}) W = f(\xi) W, \quad ' = \frac{d}{d\xi}, \tag{3.33}$$

or

$$W'' - (a^4 + \frac{\nu^2 - 1/4}{\xi^2}) W = f(\xi) W, \quad ' = \frac{d}{d\xi}. \tag{3.34}$$

The solutions of the corresponding "unperturbed" equations in this case are expressed in terms of Bessel functions, for which we abide by the notation of Watson's treatise [9]. We investigate Eqs. (3.33) and (3.34) on the finite interval $(0, \xi_0]$, $\xi_0 > 0$; for the two linearly independent solutions of Eq. (3.33) with $f \equiv 0$ we adopt the functions $\sqrt{\xi}\, J_\nu(a^2\xi)$ and $\sqrt{\xi}\, Y_\nu(a^2\xi)$, while for the "unperturbed" equation (3.34) we adopt the functions $\sqrt{\xi}\, I_\nu(a^2\xi)$ and $\sqrt{\xi}\, K_\nu(a^2\xi)$. The Wronskians of these pairs are

$$\mathfrak{W}[\sqrt{\xi}\, J_\nu(a^2\xi), \sqrt{\xi}\, Y_\nu(a^2\xi)] = \frac{2}{\pi}, \tag{3.35}$$

$$\mathfrak{W}[\sqrt{\xi}\, I_\nu(a^2\xi), \sqrt{\xi}\, K_\nu(a^2\xi)] = -1. \tag{3.36}$$

Also, we deduce the following inequalities at once from the well-known asymptotic expansions of Bessel functions of a large argument.

Lemma 1. For every fixed $\nu \ge 0$ there exists a constant c_ν such that for $\xi \ge 0$ and $a \ge 1$

$$\sqrt{\xi}\, |J_\nu(a^2\xi)| \le c_\nu \frac{\sqrt{\xi}}{(1+(a^2\xi)^{-\nu})(1+a\sqrt{\xi})}, \tag{3.37}$$

$$0 \leq \sqrt{\xi}\, I_\nu(a^2\xi) \leq c_\nu \frac{\sqrt{\xi}\, e^{a^2\xi}}{(1+(a^2\xi)^{-\nu})(1+a\sqrt{\xi})}. \tag{3.38}$$

Lemma 2. For every fixed $\nu > 0$ there exists a constant c_ν such that for and $a \geq 1$

$$\sqrt{\xi}\, |Y_\nu(a^2\xi)| \leq c_\nu (1+(a^2\xi)^{-\nu}) \frac{\sqrt{\xi}}{1+a\sqrt{\xi}}, \tag{3.39}$$

$$0 \leq \sqrt{\xi}\, K_\nu(a^2\xi) \leq c_\nu (1+(a^2\xi)^{-\nu}) \frac{\sqrt{\xi}\, e^{-a^2\xi}}{1+a\sqrt{\xi}}. \tag{3.40}$$

Lemma 3. There is a constant c_0 such that for all $\xi > 0$ and $a \geq 1$

$$\sqrt{\xi}\, |Y_0(a^2\xi)| \leq c_0 \frac{1+|\ln(a^2\xi)|}{\ln(2+a^2\xi)} \frac{\sqrt{\xi}}{1+a\sqrt{\xi}}, \tag{3.41}$$

$$0 \leq \sqrt{\xi}\, K_0(a^2\xi) \leq c_0 \frac{1+|\ln(a^2\xi)|}{\ln(2+a^2\xi)} \frac{\sqrt{\xi}\, e^{-a^2\xi}}{1+a\sqrt{\xi}}. \tag{3.42}$$

Now the obvious application of Theorem 8 to Eq. (3.33) or (3.34) yields the following inequalities.

Theorem 10. In Eq. (3.33) let $a \geq 1$, $\nu \geq 0$ and let the function $\xi(1+|\ln\xi|)\,|f(\xi)|$ be integrable in the interval $[0, \xi_0]$. Then there exist solutions W_1 and W_2 of this equation such that for $0 < \xi \leq \xi_0$

$$|W_1 - \sqrt{\xi}\, J_\nu(a^2\xi)| \leq \frac{c_\nu \sqrt{\xi}\,(e^{F(\xi)} - 1)}{(1+(a^2\xi)^{-\nu})(1+a\sqrt{\xi})}, \tag{3.43}$$

$$|W_2 - \sqrt{\xi}\, Y_\nu(a^2\xi)| \leq \frac{c_\nu \sqrt{\xi}\,(1+(a^2\xi)^{-\nu})}{1+a\sqrt{\xi}} \frac{1+|\ln(a^2\xi)|}{\ln(2+a^2\xi)} \left(e^{F(\xi_0)-F(\xi)} - 1\right), \tag{3.44}$$

where

$$F(\xi) = \begin{cases} \pi c_\nu^2 \displaystyle\int_0^\xi \frac{t|f(t)|}{(1+a\sqrt{t})^2}\, dt, & \nu > 0; \quad (3.45) \\[2ex] \pi c_0^2 \displaystyle\int_0^\xi \frac{1+|\ln(a^2t)|}{\ln(2+a^2t)} \frac{t|f(t)|}{(1+a\sqrt{t})^2}\, dt, & \nu = 0. \quad (3.46) \end{cases}$$

Theorem 11. In Eq. (3.34) let the parameters and the function f satisfy the same conditions as in the preceding theorem. Then there exist solutions V_1 and V_2 of this equation such that for $0 < \xi \leq \xi_0$.

$$|V_1 - \sqrt{\xi}\, I_\nu(a^2\xi)| \leq \frac{c_\nu \sqrt{\xi}\, e^{a^2\xi}(e^{F(\xi)} - 1)}{(1+(a^2\xi)^{-\nu})(1+a\sqrt{\xi})}, \tag{3.47}$$

$$|V_2 - \sqrt{\xi}\, K_\nu(a^2\xi)| \leq \frac{c_\nu \sqrt{\xi}\,(1+(a^2\xi)^{-\nu})(1+|\ln(a^2\xi)|)}{(1+a\sqrt{\xi})\ln(2+a^2\xi)} e^{-a^2\xi}\left(e^{F(\xi_0)-F(\xi)} - 1\right), \tag{3.48}$$

where $F(\xi)$ is determined by Eqs. (3.45)-(3.46).

3.5. Whittaker Functions. We next consider the derivation of inequalities for the "perturbed" Whittaker equation. These inequalities are needed when an equation of the type (3.1) is investigated in an interval containing both a regular singular point and a turning point such that the position of the turning point depends on a parameter and for a certain value of the latter the turning point coalesces with the singular point. In this case the uniform approximation of the solutions of (3.1) by Bessel functions becomes untenable, and it is necessary to adopt an equation similar to the equation for a confluent hypergeometric function as our standard equation. The analysis of the eigenfunctions of the boundary-value problem (1.3) is facilitated by specifying this equation in the form

$$\frac{d^2w}{dx^2} + \left(\pm\frac{1}{4}a^4 \pm \frac{\gamma^2 a^4}{x} + \frac{\nu^2-1}{4x^2}\right)w = f(x)\,w. \tag{3.49}$$

In this and the next two subsections we describe certain facts that we shall need later on from the theory of the corresponding unperturbed equation. We use the standard Whittaker form

$$\frac{d^2w}{dz^2} + \left(-\frac{1}{4} + \frac{\varkappa}{z} + \frac{1/4-\mu^2}{z^2}\right)w = 0\,. \tag{3.50}$$

In this equation z is the complex variable, and $\varkappa$ and μ are parameters. For our analysis we require estimates of the solutions of this equation in two special cases: $\mu>0$ for nonnegative real $\varkappa$ and z and $\mu=0$ for purely imaginary $\varkappa$ and z.

The two Whittaker functions representing solutions of Eq. (3.50) are determined as follows (see [6], pp. 163 and 175). We put

$$M_{\varkappa,\mu}(z) = z^{\mu+\frac{1}{2}} e^{-z/2} \sum_{n=0}^{\infty} \frac{\Gamma(2\mu+1)\Gamma(\mu-\varkappa+1/2+n)}{\Gamma(\mu-\varkappa+\frac{1}{2})\Gamma(2\mu+1+n)n!} z^n \tag{3.51}$$

[where $z^{\mu+\frac{1}{2}} = |z|^{\mu+1/2} \exp(i(\mu+\frac{1}{2})\arg z$] and define $W_{\varkappa,\mu}(z)$ by the equation

$$W_{\varkappa,\mu}(z) = \frac{\Gamma(-2\mu)}{\Gamma(-\mu-\varkappa+1/2)} M_{\varkappa,\mu}(z) + \frac{\Gamma(2\mu)}{\Gamma(\mu-\varkappa+1/2)} M_{\varkappa,-\mu}(z), \tag{3.52}$$

in which the limiting value is taken for the right-hand side if 2μ is integer or zero-valued.

Inasmuch as Eq. (3.50) is invariant under replacement of μ by $-\mu$ or under simultaneous changes of sign of $\varkappa$ and z, it has as solutions, along with $M_{\varkappa,\mu}(z)$ and $W_{\varkappa,\mu}(z)$, the functions $M_{\varkappa,-\mu}(z)$, $W_{\varkappa,-\mu}(z)$, $M_{-\varkappa,\mu}(ze^{\pm i\pi})$, $M_{-\varkappa,-\mu}(ze^{\pm i\pi})$, $W_{-\varkappa,\mu}(ze^{\pm i\pi})$, and $W_{-\varkappa,-\mu}(ze^{\pm i\pi})$. Any three of these twelve solutions are linearly dependent; in particular,*

$$M_{\varkappa,\mu}(z) = \frac{e^{-i\varkappa\pi}\Gamma(2\mu+1)}{\Gamma(\mu-\varkappa+1/2)} W_{-\varkappa,\mu}(ze^{-i\pi}) + \frac{e^{(\mu-\varkappa+1/2)i\pi}\Gamma(2\mu+1)}{\Gamma(\mu+\varkappa+1/2)} W_{\varkappa,\mu}(z). \tag{3.53}$$

Moreover, the function $z^{-\mu-1/2} M_{\varkappa,\mu}(z)$ (which is obviously a single-valued function of z), according to the Kummer transformation ([6], p. 163), is left invariant by simultaneous change of sign of $\varkappa$ and z:

$$z^{-\mu-1/2} M_{\varkappa,\mu}(z) = (-z)^{-\mu-1/2} M_{-\varkappa,\mu}(-z). \tag{3.54}$$

Finally, it follows directly from the definition (3.52) that $W_{\varkappa,\mu}(z)$ is an even function of the parameter μ.

3.6. Asymptotic Expansions of the Whittaker Functions. The following equation ([6], p. 170) holds for the function $W_{\varkappa,\mu}(z)$ when $|z|\to$ (and fixed values of the parameters $\varkappa$ and μ are given):

*In [6], p. 176 [Russian edition], there are two misprints in this equation.

$$W_{\varkappa,\mu}(z) = z^{\varkappa} e^{-z/2}\left[1 + O\left(\frac{1+|\varkappa|^2}{z}\right)\right], \quad |\arg z| \leq \frac{3\pi}{2}. \tag{3.55}$$

Along with relations (3.53) this gives the asymptotic behavior as $|z| \to \infty$ of the function $M_{\varkappa,\mu}(z)$, when $-\frac{\pi}{2} < \arg z < \frac{3\pi}{2}$, and the Kummer transformation can be used to obtain from this the asymptotic behavior of $M_{\varkappa,\mu}(z)$ for $-\frac{3\pi}{2} < z < \frac{\pi}{2}$ as well.

The asymptotic behavior of the Whittaker functions as $|\varkappa| \to \infty$ is far more complicated. Many authors have investigated it; the most thorough-going analyses are given by Erdelyi and Swenson in [20] and by Skovgaard in [27]. In [20] Erdelyi and Swenson obtain uniform asymptotic equations for the functions $M_{\varkappa,\mu}(4\varkappa z)$ and $W_{\varkappa,\mu}(4\varkappa z)$ for arbitrary complex $\varkappa$, $|\varkappa| \gg \varkappa_0$ and real $\varkappa$. In the case of large positive $\varkappa$, however, only the first and last of the fundamental results of [20], i.e., Eqs. (10.1)-(10.4) on page 31, are valid.*

Complete asymptotic expansions of the Whittaker functions are obtained in [27]; we now give from this study the results that we shall require below, retaining only the first term of the corresponding expansions and somewhat modifying the notation.

Let the functions $\eta(z)$ and $\xi(z)$ be defined by the equations

$$\eta(z) = \int_0^z \sqrt{\frac{1}{t} - 1}\, dt, \quad 0 \leq z < 1, \tag{3.56}$$

$$(z) = \begin{cases} -\left(\frac{3}{2}\int_z^1 \sqrt{\frac{1}{t}-1}\, dt\right)^{2/3}, & 0 < z \leq 1; \quad (3.57) \\ \left(\frac{3}{2}\int_1^z \sqrt{1-\frac{1}{t}}\, dt\right)^{2/3}, & 1 \leq z < +\infty. \quad (3.58) \end{cases}$$

The asymptotic equations for the Whittaker functions have the following form in this notation.

<u>Theorem 12.</u> Let $\varkappa \to +\infty$ and $\mu > 0$ be fixed. Then for $0 \leq z \leq 3/4$ and for

$$M_{\varkappa,\mu}(4\varkappa z) = 2^{1/2}\Gamma(2\mu+1)\varkappa^{-\mu+1/2}\left(\frac{d\eta}{dz}\right)^{-1/2}\left(\sqrt{\eta}\, J_{2\mu}(2\varkappa\eta(z)) + O\left(\frac{\varkappa^{2\mu-1}\eta^{2\mu+3/2}}{(1+\sqrt{\varkappa\eta})(1+(\varkappa\eta)^{2\mu})}\right)\right), \tag{3.59}$$

and for $1/4 \leq z < +\infty$

$$M_{\varkappa,\mu}(4\varkappa z) = \pi^{1/2}\Gamma(2\mu+1)(2\varkappa)^{1/6}\left(\frac{d\xi}{dz}\right)^{-1/2}\left(\left(\sin\pi(\varkappa-\mu) + O\left(\frac{1}{\varkappa}\right)\right)U_1(\xi(z)) + \frac{2\varkappa^{-\varkappa}e^{\varkappa}}{\Gamma(\mu-\varkappa+1/2)}\left(1+O\left(\frac{1}{\varkappa}\right)\right)U_2(\xi(z))\right), \tag{3.60}$$

where

$$U_1(\xi) = Ai\left((2\varkappa)^{2/3}\xi\right) + O\left(\frac{|\exp(-4/3\,\varkappa\xi^{3/2})|}{\varkappa(1+|\xi|^{3/2})(1+\varkappa^{1/6}|\xi|^{1/4})}\right), \tag{3.61}$$

*Equation (10.2) is derived with the aid of the Stirling expansion of $\Gamma(\mu - \varkappa + 1/2)$ which becomes meaningless when $\mu - \varkappa + 1/2$ is equal to a negative integer. Also in this case Eq. (10.3) is devoid of significance, as it only gives the estimate $O(e^{2\varkappa z})$ for a function that decays exponentially as $z \to \infty$.

$$U_2(\xi) = Bi\left((2\varkappa)^{2/3}\xi\right) + O\left(\frac{\left|\exp\left(\frac{4}{3}\varkappa\xi^{3/2}\right)\right|}{\varkappa\left(1+\varkappa^{1/6}|\xi|^{1/4}\right)}\right). \tag{3.62}$$

Theorem 13.* Under the same conditions as in Theorem 12, we have for

$$W_{\varkappa,\mu}(4\varkappa z) = 2^{3/2}\varkappa^{\varkappa+\frac{1}{2}}e^{-\varkappa}\left(\sin\pi(\varkappa-\mu)+O\left(\tfrac{1}{\varkappa}\right)\right)\left(\frac{d\eta}{dz}\right)^{-1/2}\left(\sqrt{\eta}\,J_{2\mu}(2\varkappa\eta)+O\left(\frac{\eta^{2\mu+3/2}\varkappa^{2\mu-1}}{(1+\sqrt{\varkappa\eta})(1+(\varkappa\eta)^{2\mu})}\right)\right)+$$

$$+\frac{(-2^{1/2})\pi\varkappa^{\mu}}{\Gamma(\mu-\varkappa+1/2)}\left(1+O\left(\tfrac{1}{\varkappa}\right)\right)\left(\frac{d\eta}{dz}\right)^{-1/2}\left(\sqrt{\eta}\,Y_{2\mu}(2\varkappa\eta)+O\left(\frac{\sqrt{\eta}\,(1+(\varkappa\eta)^{-2\mu}}{\varkappa\,(1+\sqrt{\varkappa\eta}}\right)\right), \tag{3.63}$$

and for $1/4 \le z < +\infty$

$$W_{\varkappa,\mu}(4\varkappa z) = 2^{7/6}\pi^{1/2}\varkappa_1^{\varkappa+1/6}\left(\frac{d\xi}{dz}\right)^{-1/2}U_1(\xi(z)). \tag{3.64}$$

We shall also need asymptotic equations for the Whittaker functions with $\mu=0$ in the case when z and $\varkappa$ assume purely imaginary values and $|\varkappa|\to\infty$ (see [20]).

Theorem 14. Let $\varkappa$ be positive and large, and let z be real. Then for $-\infty < z \le 3/4$

$$M_{i\varkappa,0}(4i\varkappa z) = e^{i\pi/4}\left(\frac{2\varkappa\eta}{\eta'}\right)^{1/2}\left(I_0(2\varkappa\eta)+O\left(\frac{\eta\,|e^{2\varkappa\eta}|}{(1+|\varkappa\eta|^{1/2})(1+|\varkappa\eta|}\right)\right), \tag{3.65}$$

$$W_{i\varkappa,0}(4i\varkappa z) = e^{i(\varkappa\ln\varkappa-\varkappa+3\pi/4)+\frac{\pi}{2}\varkappa}\left(\frac{\pi\varkappa\eta}{\eta'}\right)^{1/2}\left(K_0(2\varkappa\eta)+O\left(\frac{|e^{-2\varkappa\eta}|\,(1+|\ln|\varkappa\eta||)}{\varkappa\,(1+|\varkappa\eta|^{1/2})\ln(2+|\varkappa\eta|)}\right)\right), \tag{3.66}$$

where the function $\eta(z)$ is defined by Eq. (3.56) for $z>0$ and it is assumed for $z<0$ that $\arg(iz)=-\pi/2$, $\arg\eta(z)=-\pi/2$, and

$$\eta(z) = -i\int_z^0\sqrt{\frac{1}{t}-1}\,dt. \tag{3.67}$$

Theorem 15. Let $\varkappa$ be positive and large; then for real values of $z \ge 1/4$

$$M_{i\varkappa,0}(4\varkappa iz) = \left(\frac{\varkappa}{2}\right)^{1/6}e^{\pi\varkappa+i\pi/4}(\xi')^{-1/2}\left(Ai\left(-(2\varkappa)^{2/3}\xi(z)\right)+O\cdot\left(\frac{\left|\exp\left(-\frac{4}{3}\varkappa(-\xi)^{3/2}\right)\right|}{\varkappa\left(1+\varkappa^{1/6}|\xi|^{1/4}\right)}\right)\right), \tag{3.68}$$

$$W_{i\varkappa,0}(4i\varkappa z) = (4\varkappa)^{1/6}\pi^{1/2}\exp\left(-\frac{\pi}{8}\varkappa+i\left(\varkappa\ln\varkappa-\varkappa+\frac{\pi}{12}\right)\right)(\xi')^{-1/2}\left(Ai\left(\left(2\varkappa e^{i\pi/2}\right)^{2/3}\xi(z)\right)+O\left(\frac{\left|\exp\left(-\frac{4}{3}i\varkappa\xi^{3/2}\right)\right|}{\varkappa\left(1+\varkappa^{1/6}|\xi|^{1/4}\right)}\right)\right), \tag{3.69}$$

where the function $\xi(z)$ is defined by Eqs. (3.57)-(3.58).

3.7. Inequalities. Here we obtain inequalities for the Whittaker functions and use them to estimate the solutions of Eq. (3.49).

We first consider the real case. We shall assume that $\varkappa \ge \mu > 0$, $z>0$ we fix the parameter μ

*The asymptotic equations of [27] for the interval $0<z\le 3/4$ become invalid when $\varkappa-\mu-1/2$ is equal to a positive integer; the corrected result is given here.

once and for all. If $\varkappa$ does not exceed an absolute constant, then as $z \to 0$ we obtain directly from (3.51) and (3.52)

$$|M_{\varkappa,\mu}(z)| \le c_1 z^{\mu+1/2}, \quad |W_{\varkappa,\mu}(z)| \le c_2 \left(\alpha(\varkappa) z^{-\mu+1/2} + z^{\mu+1/2}\right), \tag{3.70}$$

where $\alpha(\varkappa) = |\cos\pi(\varkappa-\mu)|$. Moreover, by virtue of (3.53), $|W_{-\varkappa,\mu}(ze^{-i\pi})| \le c_3 z^{-\mu+1/2}$. The constant $c_1, c_2, \ldots$ are independent of $\varkappa$ and z (but can depend on μ). Then, by virtue of (3.55), as $z \to +\infty$

$$|W_{-\varkappa,\mu}(ze^{-i\pi})| \le c_4 z^{-\varkappa} e^{z/2}, \quad |W_{\varkappa,\mu}(z)| \le c_5 z^{\varkappa} e^{-z/2}, \tag{3.71}$$

$$|M_{\varkappa,\mu}(z)| \le c_6 \left(\alpha(\varkappa) z^{-\varkappa} e^{z/2} \Gamma(\varkappa-\mu+1/2) + \frac{z^{\varkappa} e^{-z/2}}{\Gamma(\varkappa+\mu+1/2)}\right). \tag{3.72}$$

We transform the latter inequalities as follows. Let $\vartheta(x) = \int_1^x \sqrt{1-\frac{1}{t}}\,dt$ for $\varkappa \ge 1$ and $\vartheta(x) = 0$ for $0 \le x \le 1$. We first show that inequalities (3.71) can be written in the form

$$|W_{-\varkappa,\mu}(ze^{-i\pi})| \le c_7 \exp\left(2\varkappa\vartheta\left(\frac{z}{4\varkappa}\right) - \varkappa\ln\varkappa + \varkappa\right),$$
$$|W_{\varkappa,\mu}(z)| \le c_8 \exp\left(-2\varkappa\vartheta\left(\frac{z}{4\varkappa}\right) + \varkappa\ln\varkappa - \varkappa\right) \tag{3.73}$$

and that they are valid in this form for arbitrarily large values of $\varkappa$, provided only that $z \ge 5\varkappa$. Thus,

$$\frac{1}{8x} \le \vartheta(x) - \left(x - \frac{1}{2}\ln 4x - \frac{1}{2}\right) \le \frac{1}{2x},$$

since

$$\vartheta(x) = x - \frac{1}{2}\ln x + \int_1^x \left(\sqrt{1-\frac{1}{t}} - 1 + \frac{1}{2t}\right) dt = x - \frac{1}{2}\ln(4x) - \frac{1}{2} + \frac{1}{4}\int_x^\infty \frac{dt}{t^2\left(1+\sqrt{1-\frac{1}{t}}\right)^2}.$$

Substituting these expressions for $\vartheta(x)$ into (3.73), we readily verify that the above inequalities are equivalent to (3.71). Also using the inequality $\Gamma(x+1/2) \le c_9 x^x e^{-x}$ (which holds for $x \ge 0$), we can write in place of (3.72) for $z \ge 5\varkappa$

$$|M_{\varkappa,\mu}(z)| \le c_{10}(1+\varkappa)^{-\mu}\left(\alpha(\varkappa) e^{2\varkappa\vartheta(z/4\varkappa)} + e^{-2\varkappa\vartheta(z/4\varkappa)}\right). \tag{3.74}$$

Now, using the asymptotic equations of Theorems 12 and 13 and the inequalities for the Airy functions, we verify at once that inequalities (3.73) and (3.74) also hold for large $\varkappa$ if $z \ge 4\varkappa(1+\delta)$ with a certain fixed $\delta > 0$. Moreover, if the right-hand sides of these inequalities are multiplied by $\varkappa^{1/6}\left(1+\varkappa^{1/6}\left|1-\frac{4\varkappa}{z}\right|^{1/4}\right)^{-1}$, the resulting inequalities will hold for all $z \ge \varkappa$. Finally, in the interval $0 < z \le \varkappa$ it follows from (3.59) that

$$|M_{\varkappa,\mu}(z)| \le c_{11}\varkappa^{-\mu+1/2}\left(\eta'\left(\frac{z}{4\varkappa}\right)\right)^{-1/2} \min\left(\varkappa^{2\mu}\left(\eta\left(\frac{z}{4\varkappa}\right)\right)^{2\mu+\frac{1}{2}}, \frac{1}{\sqrt{\varkappa}}\right). \tag{3.75}$$

Recognizing that $\sqrt{x} \le \eta(x) \le 2\sqrt{x}$ for $0 \le x \le 1/4$ and $\eta' \sim \frac{1}{\sqrt{x}}$, we can write in place of (3.75)

$$|M_{\varkappa,\mu}(z)| \le c_{12}\varkappa^{-\mu}\min\left(\varkappa^{\mu} z^{\mu+1/2}, \left(\frac{z}{\varkappa}\right)^{1/4}\right) = c_{13}\varkappa^{-\mu}\left(\varkappa^{-\mu} z^{-\mu-1/2} + \left|1-\frac{4\varkappa}{z}\right|^{1/4}\right)^{-1}. \tag{3.76}$$

Combining these inequalities with (3.74) (in which $\vartheta = 0$ when $z \le 4\varkappa$), we arrive at the following result.

Lemma 4. For every fixed $\mu > 0$ there is an absolute constant $c(\mu)$ such that for any $\varkappa \geq \mu$ and any $z > 0$

$$|M_{\varkappa,\mu}(z)| \leq c(\mu)\varkappa^{-\mu} G(z)\left(|\cos\pi(\varkappa-\mu)| e^{2\varkappa\vartheta\left(\frac{z}{4\varkappa}\right)} + e^{-2\varkappa\vartheta\left(\frac{z}{4\varkappa}\right)}\right), \tag{3.77}$$

where

$$G(z) = \left(\varkappa^{-1/6} + \varkappa^{-\mu} z^{-\mu-\frac{1}{2}} + \left|1 - \frac{4\varkappa}{z}\right|^{1/4}\right)^{-1}, \tag{3.78}$$

and the function $\vartheta(x)$ is equal to $\int_1^x \left(1-\frac{1}{t}\right)^{1/2} dt$ for $x \geq 1$ and is equal to 0 for $x < 1$.

As is evident from the technique by which this lemma was deduced, inequality (3.77) is sharp and cannot be improved.

The analogous inequality for the function $W_{\varkappa,\mu}(z)$ is proved by exactly the same reasoning; then the estimate for $W_{-\varkappa,\mu}(ze^{-i\pi})$ follows at once from (3.53). We have the following consequence.

Lemma 5. Under the notation and conditions of the preceding lemma

$$|W_{\varkappa,\mu}(z)| \leq c(\mu)\left(1 + \frac{|\cos\pi(\varkappa-\mu)|}{\varkappa^{\mu} z^{\mu}}\right) G(z) \exp\left(-2\varkappa\vartheta\left(\frac{z}{4\varkappa}\right) + \varkappa\ln\varkappa - \varkappa\right), \tag{3.79}$$

$$|W_{-\varkappa,\mu}(ze^{-i\pi})| \leq c(\mu)\left(1 + (\varkappa z)^{-\mu}\right) G(z) \exp\left(2\varkappa\vartheta\left(\frac{z}{4\varkappa}\right) - \varkappa\ln\varkappa + \varkappa\right). \tag{3.80}$$

3.8. Approximation by Whittaker Functions (Real Case). Consider the equation

$$\frac{d^2W}{dx^2} + \left(-\frac{a^4}{4} + \frac{\gamma^2 a^2}{x} - \frac{(\nu^2 - 1/4)}{x^2}\right) W = f(x) W, \tag{3.81}$$

in which we assume $a \geq 1$, $\gamma a \geq \nu/2 > 0$; we also postulate that the function $x|f(x)|$ is integrable in a certain interval $[0, x_0]$, $x_0 > 0$

The following functions are solutions of the unperturbed equation:

$$w_1 = M_{\varkappa,\nu/2}(a^2 x), \quad w_2 = W_{\varkappa,\nu/2}(a^2 x); \quad w_3 = W_{-\varkappa,\nu/2}(a^2 x e^{-i\pi}),$$

where $\varkappa = a^2\gamma^2$ The Wronskians of these pairs are easily computed by means of (3.51)-(3.53). We thereby obtain

$$\mathfrak{B}[w_1, w_2] \equiv w_1 \frac{dw_2}{dx} - w_2 \frac{dw_1}{dx} = -\frac{a^2\Gamma(\nu+1)}{\Gamma(\nu/2 - \varkappa + 1/2)}, \tag{3.82}$$

$$\mathfrak{B}[w_1, w_3] = \frac{a^2\Gamma(\nu+1)e^{\frac{i\pi}{2}(\nu+1)}}{\Gamma(\nu/2 + \varkappa + 1/2)}, \quad \mathfrak{B}[w_2, w_3] = a^2 e^{i\pi\varkappa}. \tag{3.83}$$

In deriving these equations we at first assume that ν is not an integer; because of continuity, however, the result also holds for integer-valued ν. We partition the interval $[0, x_0]$ into subintervals $(0, 4\gamma^2)$ and $(4\gamma^2, x_0)$ (assuming that $x_0 > 4\gamma^2$). Choosing the functions w_1 and w_3 in the first subinterval and w_2 and w_3 in the second subinterval as linearly independent solutions of the unperturbed equation and applying Theorem 8 and the inequalities of Lemma 4 and 5 in straightforward fashion, we arrive at the following.

Theorem 16. There exist solutions W_1, W_2, and W_3 of Eq. (3.81) such that for $0 \leq x \leq 4\gamma^2$

$$|W_1 - M_{\varkappa,\nu/2}(a^2x)| \leq c(\nu)\varkappa^{-\nu/2} G(a^2x)\left(\exp\left(\frac{c^2(\nu)}{a^2}\int_0^x (1+(a\varkappa)^{-\nu} t^{-\nu/2}) G^2(a^2t)|f(t)|dt\right)-1\right), \tag{3.84}$$

and for $4\gamma^2 \leq x \leq x_0$

$$|W_2 - W_{\varkappa,\nu/2}(a^2x)| \exp\left(2\varkappa\vartheta\left(\frac{x}{4\gamma^2}\right)+\varkappa-\varkappa\ln\varkappa\right) \leq c(\nu)\, G(a^2x)\left(\exp\left(\frac{c^2(\nu)}{a^2}\int_x^{x_0} G^2(a^2t)|f(t)|dt\right)-1\right), \tag{3.85}$$

$$|W_3 - W_{-\varkappa,\nu/2}(a^2xe^{-i\pi})| \exp\left(-2\varkappa\vartheta\left(\frac{x}{4\gamma^2}\right)-\varkappa+\varkappa\ln\varkappa\right) \leq c(\nu) G(a^2x)\left(\exp\left(\frac{c^2(\nu)}{a^2}\int_{4\gamma^2}^x G^2(a^2t)|f(t)|dt\right)-1\right), \tag{3.86}$$

where $c(\nu)$ is a constant depending only on ν.

3.9. Approximation by Whittaker Functions (Nonreal Case). We complete the present section with an estimate of that solution of the equation

$$\frac{d^2w}{dx^2} + \left(\frac{a^4}{4} - \frac{\gamma^2a^4}{x} - \frac{1}{4x^2}\right) w = f(x)\, w \tag{3.87}$$

for which $x^{-1/2}|w(x)|$ remains finite as $x \to 0$. We assume here that the function $|x f(x)|(1+|\ln x|)$ is integrable in an interval of the type $|x| \leq x_0$, $x_0 > 0$, and $a \geq 1$, $\gamma \geq 0$. Before formulating the appropriate result, we write out inequalities for the functions $M_{i\varkappa,0}(ia^2x)$ and $W_{i\varkappa,0}(ia^2x)$, $\varkappa = \gamma^2a^2$, satisfying Eq. (3.87) for $f \equiv 0$

Lemma 6. Let $\tilde{\vartheta}(x)$ be defined by the equations

$$\tilde{\vartheta}(x) = \begin{cases} 0, & -\infty < x \leq 0; \\ \int_0^x \sqrt{\frac{1}{t}-1}\, dt, & 0 \leq x \leq 1; \\ \frac{\pi}{2}, & x > 1. \end{cases} \tag{3.88}$$

Let $\varkappa > 0$, $a \geq 1$ then there is an absolute constant b_0 such that for $-\infty < x < +\infty$

$$|M_{i\varkappa,0}(ia^2x)| \leq b_0\, \beta_1(x) \exp\left(2\varkappa\tilde{\vartheta}\left(\frac{x}{4\gamma^2}\right)\right), \tag{3.89}$$

$$|W_{i\varkappa,0}(ia^2x)| \leq b_0\, \beta_2(x) \exp\left(-2\varkappa\tilde{\vartheta}\left(\frac{x}{4\gamma^2}\right)+\frac{\pi}{2}\varkappa\right), \tag{3.90}$$

where

$$\beta_1(x) = \left((a^2|x|)^{-1/2} + (1+\varkappa)^{-1/6} + \left|1-\frac{4\gamma^2}{x}\right|^{1/4}\right)^{-1}, \tag{3.91}$$

$$\beta_2(x) = \left((a^2|x|)^{-1/2}(1+|\ln|a^2x||) + (1+\varkappa)^{-1/6} + \left|1-\frac{4\gamma^2}{x}\right|^{1/4}\right)^{-1}. \tag{3.92}$$

For the proof of the lemma in the case of large $\varkappa$ it suffices to compare the asymptotic equations of Theorems 15 and 14 with the behavior of the right-hand sides of (3.89) and (3.90); for small values of $\varkappa$ these inequalities are directly implied by (3.51)-(3.55).

The following is a direct consequence of this lemma and Theorem 8.

Theorem 17. There exists a solution $W(x)$ of Eq. (3.87) such that

$$\lim_{x\to 0} x^{-1/2} W(x) = a$$

and for any x, $|x|\leq x_0$, this solution satisfies the inequality

$$|W(x) - M_{i\varkappa,0}(ia^2x)| \leq b_0\, \beta_1(x)\, e^{2\varkappa\tilde{\vartheta}\left(\frac{x}{4\gamma^2}\right)}\left(\exp\left(\frac{b_0^2}{a^2}|F(x)| - 1\right)\right), \tag{3.93}$$

in which, in the notation of Lemma 6,

$$F(x) = \int_0^x \beta_1(t)\beta_2(t)|f(t)|\,dt. \tag{3.94}$$

§ 4. Asymptotic Equations for the First Eigenfunctions

4.1. Preliminary Remarks. We recall that the eigenfunctions $\psi_n(x,a)$, $n=0,1,\ldots$, of the integral equation

$$\mu\,\psi(x) = \int_0^a \sqrt{x\xi}\, J_\nu(x\xi)\,\psi(\xi)\,d\xi, \quad \nu > \tfrac{1}{2}$$

are also eigenfunctions of the boundary-value problem

$$\frac{d}{dx}(x^2 - a^2)\frac{d}{dx}\psi + a^2\left(x^2 + \frac{\nu^2 - 1/4}{x^2}\right)\psi = \tilde{\lambda}\psi, \quad |\psi(0)|, |\psi(a)| < \infty, \tag{4.1}$$

in which the spectral parameter may be regarded as positive and satisfying the inequality $\tilde{\lambda} > \sqrt{4\nu^2 - 1}\, a^2$.

In § 3 we deduced a series of inequalities for the solutions of certain second-order differential equations; we shall now use those inequalities to obtain asymptotic equations for the first eigenfunctions of the boundary-value problem (4.1). We regard as the "first" eigenfunctions those for which their corresponding eigenvalues $\tilde{\lambda}_n$ satisfy the inequality $\tilde{\lambda}_n \leq a^4/2$. We shall witness later on that this inequality is clearly fulfilled when $n \leq A_0 a^2$, $A_0 = \frac{1}{2\pi}\int_0^1 \sqrt{\frac{1-s^2}{1-s^{1/2}}}\,ds$. Consequently, every eigenfunction with a fixed order number, given sufficiently large a, falls into the "first" category. Thus, it suffices to pass to the limit as $a \to \infty$ in order to estimate the "first" eigenfunctions of the boundary-value problem (4.1); this passage to the limit will be effected in § 5.

The procedure used to obtain the asymptotic equations for the eigenfunctions of problem (4.1) may be outlined as follows. We first derive the asymptotic behavior of a solution of (4.1) finite at zero; this asymptotic representation will be applicable for $0 \leq x \leq \frac{\sqrt{\tilde{\lambda}}}{a}$. In the intervals $\left(\frac{\sqrt{\tilde{\lambda}}}{a}, a(1-\delta)\right]$ and $(a(1-\delta), a)$ (here and throughout the entire ensuing discussion we denote by $\delta > 0$ a fixed sufficiently small number) we then obtain asymptotic equations for two linearly independent solutions and, choosing an appropriate combination of these solutions, find an asymptotic representation on the entire interval $(0, a)$ for the solution finite at zero. In the interval $(a(1-\delta), a)$ one of the indicated linearly independent solutions remains finite as $x \to a$ while the second grows without limit. Setting the coefficient of the growth solution equal to zero, we obtain secular equations for the eigenvalues of the boundary-value problem (4.1). This done, we have only to normalize the resulting solutions so as to meet the condition

$$\int_0^a \psi_n^2(x,a)\,dx = 1.$$

4.2. Transformation of the Equation. It will be to our advantage below to transform Eq. (4.1). We seek ax instead of x and write instead of the parameter $\tilde{\lambda}$ the new parameter $a^4\lambda^2$, invoking the substitution

$$\psi(ax, a) = \begin{cases} \dfrac{w(x)}{\sqrt{1-x^2}}, & 0 \le x < 1; \\ \dfrac{\tilde{w}(x)}{\sqrt{x^2-1}}, & 1 < x < \infty. \end{cases}$$

It is readily verified that both $w(x)$ and $\tilde{w}(x)$ satisfy the equation

$$w''(x) + \left(a^4 \frac{\lambda^2 - x^2}{1-x^2} + \frac{1/4 - \nu^2}{x^2(1-x^2)} + \frac{1}{(1-x^2)^2}\right) w = 0. \tag{4.2}$$

In the interval $0 \le x \le 1-\delta$ (it may be assumed for definiteness that $2^{-7} \le \delta \le 2^{-6}$) we define the function $\xi(x, \lambda)$ by the equation

$$\frac{4\gamma^2 - \xi}{4\xi}\left(\frac{d\xi}{dx}\right)^2 = \frac{\lambda^2 - x^2}{1-x^2} \tag{4.3}$$

and the auxiliary conditions

$$\frac{d\xi(x,\lambda)}{dx} > 0, \quad \xi(0,\lambda) = 0, \quad \xi(\lambda,\lambda) = 4\gamma^2. \tag{4.4}$$

These conditions ensure continuity of the function $\frac{d\xi}{dx}$ at $x=0$ and $x=\lambda$, the last condition simply representing the equation for the quantity γ^2. Integrating (4.3), we find that this condition is equivalent to the equation

$$\gamma^2 = \frac{1}{\pi}\int_0^{\lambda} \sqrt{\frac{\lambda^2 - s^2}{1-s^2}}\, ds = \frac{\lambda^2}{\pi}\int_0^1 \sqrt{\frac{1-s^2}{1-\lambda^2 s^2}}\, ds. \tag{4.5}$$

We now introduce into (4.2) the new independent variable $\xi = \xi(x, \lambda)$ and change the unknown function by setting $w(x) = \left(\frac{d\xi}{dx}\right)^{-1/2} W(\xi)$. A direct calculation shows that $W(\xi)$ satisfies the equation

$$\frac{d^2 W}{d\xi^2} + \left(-\frac{a^4}{4} + \frac{\gamma^2 a^4}{\xi} + \frac{1-\nu^2}{4\xi^2}\right) W = f_1(\xi) W, \tag{4.6}$$

in which

$$f_1(\xi) = \left(\frac{1}{2}\{\xi, x\} - \frac{1}{(1-x^2)^2}\right)\left(\frac{dx}{d\xi}\right)^2 + \frac{1-\nu^2}{4}\left(\frac{1}{\xi^2} - \frac{1}{x^2(1-x^2)}\left(\frac{dx}{d\xi}\right)^2\right). \tag{4.7}$$

Next we let*

$$\eta(x) = \begin{cases} -\displaystyle\int_x^1 \sqrt{\frac{s^2 - \lambda^2}{1-s^2}}\, ds, & \lambda \le x \le 1; \\ \displaystyle\int_1^x \sqrt{\frac{s^2 - \lambda^2}{s^2 - 1}}\, ds, & 1 \le x < \infty \end{cases} \tag{4.8}$$

---

*This integral, as well as many others encountered below, can be expressed in terms of standard elliptic integrals. We shall not do so, because such a reduction does not shorten the notation and would render the equations physically less transparent.

(here all the roots are positive in the square). Setting

$$w(x) = \left(\frac{d\eta}{dx}\right)^{-1/2} V(\eta)$$

in Eq. (4.2), we obtain the following equation for the function $x<1$ for $V(\eta)$:

$$\frac{d^2V}{d\eta^2} + \left(-a^4 + \frac{1}{4\eta^2}\right)V = f_2(\eta)\,V, \tag{4.9}$$

in which

$$f_2(\eta) = \left(\frac{\nu^2 - 1/4}{x^2(1-x^2)} + \frac{1}{2}\{\eta, x\} - \frac{1}{(1-x^2)^2}\right)\left(\frac{dx}{d\eta}\right)^2 + \frac{1}{4\eta^2}. \tag{4.10}$$

For $x>1$ given the same function $f_2(\eta)$, the function $V(\eta)$ satisfies the equation

$$\frac{d^2V}{d\eta^2} + \left(a^4 + \frac{1}{4\eta^2}\right)V = f_2(\eta)\,V. \tag{4.11}$$

4.3. Estimates of the Solutions of Eq. (4.2) for $0 \le x \le 1-\delta$. We next estimate the function $f_1(\xi)$ in Eq. (4.6) and verify the following.

Lemma 7. There is a constant c_0 independent of a and λ such that for $0 \le \lambda \le 1/\sqrt{2}$ and $0 < x \le 1-\delta$

$$|f_1(\xi)| \le \frac{c_0}{\xi}. \tag{4.12}$$

If inequality (4.12) were already proved, it would suffice to use the result of Theorem 16 to obtain estimates of the solutions of Eq. (4.6). It is readily confirmed that the integral in the exponent of (3.84) (with f replaced by f_1) is estimated by the quantity $O(a^{\nu}\varkappa^{\nu/2}\xi^{\nu/2+1})$ for $0 \le \xi \le 1/a^2\varkappa$, and by the quantity $O\left(\frac{\sqrt{\xi}}{\gamma}\right)$ for $\frac{1}{a^2\varkappa} \le \xi \le 4\gamma^2$. The other integrals involved in the inequalities of Theorem 16 are estimated analogously; substituting the corresponding estimates, we immediately deduce the following result.

Lemma 8. There exist solutions W_1, W_2, and W_3 of Eq. (4.2) such that for

$$W_1 = \left(\frac{d\xi}{dx}\right)^{-1/2}\left(M_{\varkappa,\nu/2}(a^2\xi(x,\lambda)) + G(a^2\xi)\,O\left(\min\left(a^{\nu}\xi^{\nu/2+1}, \left(\frac{\xi}{a^2\varkappa}\right)^{1/2}\right)\right)\right), \tag{4.13}$$

and for $\lambda \le x \le 1-\delta$

$$W_2 = \left(\frac{d\xi}{dx}\right)^{-1/2}\left(W_{\varkappa,\nu/2}(a^2\xi) + G(a^2\xi)\,e^{-2\varkappa\vartheta\left(\frac{\xi}{4\gamma^2}\right) + \varkappa\ln\varkappa - \varkappa}\,O\left(\frac{1}{a^2}\ln\frac{\xi}{4\gamma^2}\right)\right), \tag{4.14}$$

$$W_3 = \left(\frac{d\xi}{dx}\right)^{-1/2}\left(W_{-\varkappa,\nu/2}(a^2\xi e^{-i\pi}) + G(a^2\xi)\,e^{2\varkappa\vartheta\left(\frac{\xi}{4\gamma^2}\right) - \varkappa\ln\varkappa + \varkappa}\,O\left(\frac{\xi(1-\delta,\lambda) - \xi(x,\lambda)}{a^2}\right)\right), \tag{4.15}$$

where $\varkappa = \gamma^2 a^2$, $\xi(x,\lambda)$ is defined by Eqs. (4.3) and the rest of the notation is consistent with that of Theorem 16.

4.4. Estimates of the Solutions of Eq. (4.2) for $x \ge \lambda+\delta$. When x is sufficiently close to unity, it follows at once from Eq. (4.8) that for $x<1$

$$1-x^2 = \frac{\eta^2}{1-\lambda^2}\left(1+O(\eta^2)\right), \quad \left(\frac{d\eta}{dx}\right)^2 = \frac{(1-\lambda^2)^2}{\eta^2}\left(1+O(\eta^2)\right), \tag{4.16}$$

and for $x>1$

$$x^2-1=\frac{\eta^2}{1-\lambda^2}(1+O(\eta^2)),\quad \left(\frac{d\eta}{dx}\right)^2=\frac{(1-\lambda^2)^2}{\eta^2}(1+O(\eta^2)). \tag{4.17}$$

Also, a straightforward calculation shows that

$$\{\eta,x\}=\frac{3}{2}\frac{x^2}{(x^2-1)^2}+\frac{x^2+\lambda^2-1}{(x^2-\lambda^2)(x^2-1)}-\frac{5}{2}\frac{x^2}{(x^2-\lambda^2)^2}. \tag{4.18}$$

For small η, therefore,

$$\frac{1}{2}\left(\frac{d\eta}{dx}\right)^{-2}\{\eta,x\}=\frac{3}{4\eta^2}+O(1). \tag{4.19}$$

Substituting the resulting estimates into (4.10), we verify the fact that the function $f_2(\eta)$ is bounded for small values of $|\eta|$. The boundedness of $f_2(\eta(x))$ for $\lambda+\delta\le x\le 1-\delta$ and $1+\delta\le x\le 1/\delta$ is obvious; but if x is sufficiently large, then

$$\frac{d\eta}{dx}=1+O\left(\frac{1}{x^2}\right),\quad \eta(x)=x+O(1),$$

and it follows directly from (4.18) and (4.10) that $f_2(\eta)=O\left(\frac{1}{\eta^2}\right)$. Inserting this estimate into (3.46), we obtain the following from Theorem 10 for the function F (with $\xi_0=+\infty$)

$$F_2(\eta)=O\left(\frac{\eta^2}{1+a^2|\eta|(1+|\eta|)}\right),\quad F(\infty)-F(\eta)=O\left(\frac{1}{a^2(1+|\eta|)}\right). \tag{4.20}$$

Consequently, by virtue of Theorems 10 and 11 in § 3 the following inequalities are applicable to the solutions of Eq. (4.2).

L e m m a 9. For $\lambda\le 1/\sqrt{2}$ there exist solutions V_1 and V_2 of Eq. (4.2) such that for $\lambda+\delta\le x<1$

$$V_1(x)=(-\eta)^{1/2}\left(\frac{d\eta}{dx}\right)^{-1/2}I_0(-a^2\eta)\left(1+O\left(\frac{\eta^2}{1+a^2|\eta|(1+|\eta|)}\right)\right), \tag{4.21}$$

$$V_2(x)=(-\eta)^{1/2}\left(\frac{d\eta}{dx}\right)^{-1/2}K_0(-a^2\eta)\left(1+O\left(\frac{1}{a^2}\right)\right), \tag{4.22}$$

and for $1\le x<+\infty$

$$V_1(x)=\eta^{1/2}\left(\frac{d\eta}{dx}\right)^{-1/2}\left(J_0(a^2\eta)+O\left(\frac{\eta^2}{(1+a\sqrt{\eta})(1+a^2\eta(1+\eta))}\right)\right). \tag{4.23}$$

4.5. Estimate of $f_1(\xi)$. In order to prove inequality (4.12) for all $\lambda\ge 0$, it is convenient to impose the substitution

$$x=\lambda\sqrt{t},\quad \xi(x,\lambda)=4\gamma^2\varphi(t),\quad 0\le t\le\frac{(1-\delta)^2}{\lambda^2}. \tag{4.24}$$

Then the function $\varphi(t)$ satisfies the equation

$$\alpha\frac{1-\varphi}{\varphi}\left(\frac{d\varphi}{dt}\right)^2=\frac{1-t}{t(1-\lambda^2 t)},\quad \alpha=\left(\frac{4\gamma^2}{\lambda}\right)^2 \tag{4.25}$$

and the conditions (guarenteeing continuity of $\frac{d\varphi}{dt}$ at $t=0$ and $t=1$),

$$\varphi(0)=0, \qquad \varphi(1)=1. \tag{4.26}$$

The function $f_1(\xi)$ is expressed in terms of φ and t as follows:

$$64\gamma^4 f_1 = (1-\nu^2)\left(\frac{1}{\varphi^2}-\frac{1}{t^2(1-\lambda^2 t)}\left(\frac{dt}{d\varphi}\right)^2\right)+\left(2\{\varphi,t\}-\frac{\lambda^2}{t(1-\lambda^2 t)^2}\right)\left(\frac{dt}{d\varphi}\right)^2. \tag{4.27}$$

In order to prove Lemma 7 we need to show that the right-hand side of (4.27) is estimated by the quantity $O(\gamma^2/\varphi)$ for $0<t\leq\frac{(1-\delta)^2}{\lambda^2}$. We can in fact demonstrate more than that, namely that this estimate is valid for negative t as well (we shall use this fact later on). For this purpose we introduce the function $F(z)$, which is defined for all real z by the equations

$$z=\begin{cases} -\int\limits_{F(z)}^{0}\sqrt{1-\frac{1}{s}}\,ds, & -\infty<F\leq 0;\\ \int\limits_{0}^{F(z)}\sqrt{1-\frac{1}{s}}\,ds, & 0<F\leq 1;\\ \frac{\pi}{2}+\int\limits_{1}^{F(z)}\sqrt{1-\frac{1}{s}}\,ds, & 1<F<\infty. \end{cases} \tag{4.28}$$

It is clear that $z(F)$ is a monotonically increasing function of F and is continuous together will all its derivatives in each of the open intervals $(-\infty,0)$, $(0,1)$, and $(1,+\infty)$ Consequently, $F(z)$ has the same properties in each of the intervals $(-\infty,0)$, $(0,\pi/2)$, and $(\pi/2,+\infty)$. With the aid of the new function $F(z)$ we write the solution of Eq. (4.25) in the form

$$\varphi(t)=F(z(t,\lambda)), \tag{4.29}$$

where

$$z(t,\lambda)=\begin{cases} -\frac{1}{\sqrt{\alpha}}\int\limits_{t}^{0}\left(\frac{s-1}{s(1-\lambda^2 s)}\right)^{1/2}ds, & -\infty<t\leq 0;\\ \frac{1}{\sqrt{\alpha}}\int\limits_{0}^{t}\left(\frac{1-s}{s(1-\lambda^2 s)}\right)^{1/2}ds, & 0\leq t\leq 1;\\ \frac{\pi}{2}+\frac{1}{\sqrt{\alpha}}\int\limits_{1}^{t}\left(\frac{s-1}{s(1-\lambda^2 s)}\right)^{1/2}ds, & 1\leq t<+\infty. \end{cases} \tag{4.30}$$

We note that $\alpha=1$ for $\lambda=0$ and $F(z(t,0))=t$ identically with respect to t. The function $\{\varphi,t\}$ therefore vanishes for $\lambda=0$ (since the Schwarzian of a linear function is identically equal to zero). Also, by virtue of (4.25) we have

$$\frac{1}{\varphi^2}-\frac{1}{t^2(1-\lambda^2 t)}\left(\frac{dt}{d\varphi}\right)^2=\frac{t(1-t)-\alpha\varphi(1-\varphi)}{t\varphi^2(1-t)}, \tag{4.31}$$

and for $\lambda=0$ this difference is also identically equal to zero. Consequently, the following equations are valid:

$$\{\varphi,t\}=\int\limits_0^{\lambda}\frac{\partial}{\partial\tau}\{F(z(t,\tau)),t\}\,d\tau, \tag{4.32}$$

$$\alpha\varphi(1-\varphi)-t(1-t)=\int\limits_0^{\lambda}\frac{\partial}{\partial\tau}\left(\alpha F(z(t,\tau))(1-F(z(t,\tau)))\right)d\tau. \tag{4.33}$$

It is easily verified that the integrand in (4.32) is estimated for large values of $|z|$ by the quantity $O\left(\frac{\tau}{z^2}\right)$ while in (4.33) the integrand is estimated by $O(\tau|z|^3)$. In order to estimate these expressions in the vicinity of the points $t=0$ and $t=1$ it suffices to use the expansions of $F(z)$ in power series in $\pm z^2$ and $(\pm(z-\pi/2))^{2/3}$, respectively, and then to substitute therein the expansions for $z(t,\tau)$ and $\frac{\partial z(t,\tau)}{\partial\tau}$ in powers of $(\pm t)^{1/2}$ and $(\pm(t-1))^{3/2}$. Omitting the necessary computations (due to their bulkiness), we present the final result.

Lemma 10. For all τ from the interval $[0, 1/\sqrt{2}]$ and all t from the interval $(-1/8, 1/8)$ the following estimates hold:

$$\left|\frac{\partial}{\partial\tau}\{F(z(t,\tau)), t\}\right| = O(\tau), \tag{4.34}$$

$$\frac{\partial}{\partial\tau}\alpha F(1-F) = O(\tau F^2(1-F)). \tag{4.35}$$

Lemma 7 follows at once from these estimates.

4.6. Asymptotic Representation on the Entire Interval $[0,1)$ for the Solution Finite at Zero. We normalize the finite-at-zero solution of Eq. (4.2) so that it will coincide with the solution W_1 of Lemma 8 in the interval $[0,\lambda)$. Then in the interval $[\lambda, 1-\delta]$ it will be a linear combination of the solutions W_2 and W_3 from the same lemma, whereas in the interval $[1/\sqrt{2}+\delta, 1)$ it will be a linear combination of the solutions V_1 and V_2 from Lemma 9.

We denote the solution thus obtained by $w(x)$; determining the coefficients of the linear combinations from the condition of continuity of w and $\frac{dw}{dx}$, we readily find

$$w(x) = \begin{cases} W_1(x), & 0 \le x \le \lambda; & (4.36) \\ \dfrac{\mathcal{D}[W_1,W_3]W_2(x) - \mathcal{D}[W_1,W_2]W_3(x)}{\mathcal{D}[W_2,W_3]}, & \lambda \le x \le 1-\delta; & (4.37) \\ A_2V_1(x) + A_1V_2(x), & \frac{1}{\sqrt{2}}+\delta \le x \le 1, & (4.38) \end{cases}$$

where $\mathcal{D}[\varphi,\psi] = \varphi\frac{d\psi}{dx} - \psi\frac{d\varphi}{dx}$ and

$$A_k = \frac{\mathcal{D}[W_1,W_3]\mathcal{D}[W_2,V_k] - \mathcal{D}[W_1,W_2]\mathcal{D}[W_3,V_k]}{\mathcal{D}[W_2,W_3]\mathcal{D}[V_1,V_2]}, \quad k=1,2. \tag{4.39}$$

We begin the calculation of the Wronskians involved here with the coefficient in front of $W_3(x)$ in (4.37). We recall that the functions $\varphi_1 = \left(\frac{d\xi}{dx}\right)^{1/2}W_1$ and $\varphi_2 = \left(\frac{d\xi}{dx}\right)^{1/2}W_2$ are solutions of the integral equation (see Sec. 3.8)

$$\varphi_1(\xi) = w_1(\xi) + \int_0^{\xi}\frac{w_1(t)w_3(\xi) - w_1(\xi)w_3(t)}{\mathcal{D}[w_1,w_3]}f(t)\varphi_1(t)\,dt, \tag{4.40}$$

$$\varphi_2(\xi) = w_2(\xi) + \int_{\xi}^{\xi_0}\frac{w_2(\xi)w_3(t) - w_2(t)w_3(\xi)}{\mathcal{D}[w_2,w_3]}f(t)\varphi_2(t)\,dt, \tag{4.41}$$

where

$$w_1 = M_{\varkappa,\nu/2}(a^2\xi), \quad w_2 = W_{\varkappa,\nu/2}(a^2\xi), \quad w_3 = W_{-\varkappa,\nu/2}(a^2\xi e^{-i\pi}).$$

Thus, if we write

$$\varphi_\kappa(\xi) = A_\kappa(\xi) w_1(\xi) + B_\kappa(\xi) w_2(\xi), \quad \kappa = 1, 2,$$

then

$$\frac{d\varphi_\kappa(\xi)}{d\xi} = A_\kappa(\xi)\frac{dw_1}{d\xi} + B_\kappa(\xi)\frac{dw_2}{d\xi}, \quad \kappa = 1, 2.$$

Consequently,

$$\mathfrak{B}[\varphi_1,\varphi_2] = A_1(\xi)A_2(\xi)\mathfrak{B}[w_1,w_2] + A_1(\xi)B_2(\xi)\mathfrak{B}[w_1,w_3] + B_1(\xi)A_2(\xi)\mathfrak{B}[w_3,w_2]. \tag{4.42}$$

The Wronskians on the right-hand side are known [see Eqs. (3.82) and (3.83)], and, since the right-hand side of (4.42) is independent of ξ, we can use the inequalities of Lemmas 5 and 4 and Theorem 16 with $\xi = 4\gamma^2$ for the functions A_1, A_2, B_1, and B_2:

$$A_1 = 1 + O(1/a^2), \quad B_1 = \varkappa^{-\nu}\Gamma(\nu/2 + \varkappa + \tfrac{1}{2})\, O(1/a^2),$$
$$A_2 = 1 + O(1/a^2), \quad B_2 = e^{2\varkappa \ln \varkappa - 2\varkappa}\, O(1/a^2\varkappa^{1/3}). \tag{4.43}$$

The first three of these equations are obtained directly; the last one requires some explanation. Using inequalities (3.85) and (3.79), we first obtain

$$|\mathfrak{D}| = \left|\frac{1}{\mathfrak{B}[w_2,w_1]}\int_{4\gamma^2}^{\xi_0} w_2(t)\varphi_2(t)f(t)\,dt\right| \leqslant \frac{c^2(\nu)}{a^2}\int_{4\gamma^2}^{\xi_0}\frac{G^2(a^2t)}{t}\exp\left(-4\varkappa\vartheta(t/4\gamma^2) + 2\varkappa\ln\varkappa - 2\varkappa\right)dt \tag{4.44}$$

(we have also made use of the estimate for $|f(t)|$ from Sec. 4.5). Inasmuch as a is large, the main contribution to the above integral originates from the neighborhood of the point $t = 4\gamma^2$. In the neighborhood of the point, $t^{-1}G^2(a^2t) = O\left(\frac{\varkappa^{2/3}}{\gamma^2}\right)$. Substituting these estimates into (4.44) and estimating the contribution of "large" values of t in the usual way, we arrive at the last of Eqs. (4.43). Then, inserting the resulting estimates into (4.42) and using the functional equation for the Γ-function, we find

$$\mathfrak{B}[\varphi_1,\varphi_2] = -\frac{a^2}{\pi}\Gamma(\nu+1)\Gamma(\varkappa - \nu/2 + 1/2)\left(1 + O\left(\frac{1}{a^2}\right)\right)\left(\cos\pi(\varkappa - \nu/2) + O\left(\frac{e^{2\varkappa\ln\varkappa - 2\varkappa}}{a^2\varkappa^{4/3}\Gamma(\varkappa + \nu/2 + 1/2)\Gamma(\varkappa - \nu/2 + 1/2)}\right) + O\left(\frac{\Gamma(\varkappa + \nu/2 + 1/2)}{a^2\varkappa^{\nu}\Gamma(\varkappa - \nu/2 + 1/2)}\right)\right). \tag{4.45}$$

Finally, taking account of the obvious identity

$$\mathfrak{B}_\varkappa[W_1,W_2](x) = \mathfrak{B}_\xi[\varphi_1,\varphi_2](\xi(x))$$

(the subscript designates the variable on which the differentiation is performed) and the well-known inequalities for the Γ-function of a positive argument, we have

$$\mathfrak{B}[W_1,W_2] = -\frac{a^2}{\pi}\Gamma(\nu+1)\Gamma(\varkappa - \nu/2 + 1/2)\left(\cos\pi(\varkappa - \nu/2) + O\left(\frac{1}{a^2}\right)\right). \tag{4.46}$$

We obtain the following equations in analogous fashion:

$$\mathfrak{B}[W_2,W_3] = e^{i\pi\varkappa}(a^2 + O(1)), \quad \mathfrak{B}[W_1,W_3] = \frac{e^{i\pi(\nu/2 + 1/2)}}{\Gamma(\varkappa + \nu/2 + 1/2)}\Gamma(\nu+1)(a^2 + O(1)),$$
$$\mathfrak{B}[V_1,V_2] = 1 + O(1/a^2). \tag{4.47}$$

The Wronskians $\mathfrak{W}[V_k, W_j]$, $k=1,2$, $j=2,3$ must be calculated by means of asymptotic expansions. In order for the result to be independent of the asymptotic equations for the derivatives of these functions, we adopt the following procedure. We have

$$W_3^{-2}\mathfrak{W}[V_1, W_3] \equiv W_3^{-2}(V_1 W_3' - W_3 V_1') = -\frac{d}{dx}\frac{V_1}{W_3}, \quad ' = \frac{d}{dx}.$$

Let x_1 and x_2 be arbitrary numbers satisfying the inequalities $\frac{1}{\sqrt{2}}+\delta \le x_1 \le 1-4\delta$, and $1-2\delta \le x_2 \le 1-\delta$. Since the Wronskian is independent of x, the latter equation yields upon integration over x between the limits from x_1 to x_2

$$\mathfrak{W}[V_1, W_3]\int_{x_1}^{x_2}\frac{dx}{W_3^2} = -\frac{V_1}{W_3}\Big|_{x_1}^{x_2}. \tag{4.48}$$

The function W_3^2, as the asymptotic equations indicate, contains the rapidly increasing factor $\exp\left(a^2\int_\lambda^x\sqrt{\frac{s^2-\lambda^2}{1-s^2}}\,ds\right)$; the main contribution to the integral therefore originates from the neighborhood of the point x_1. Repeating the standard considerations used in the Laplace method, we obtain from (4.48)

$$\mathfrak{W}[V_1, W_3] = 2a^2\sqrt{\frac{x_1^2-\lambda^2}{1-x_1^2}}\, V_1(x_1)\, W_3(x_1)\left(1+O\left(\min\left(\frac{\varkappa^2+1}{a^2}, \frac{1}{\varkappa}\right)\right)\right). \tag{4.49}$$

In order to show that (4.49) is a result of (4.48) and then to calculate the right-hand side of (4.49), we write out the asymptotic equations for the functions V_k and W_j. In the interval $x_1 \le x \le x_2$ we have

$$V_1 = \frac{1}{a}(2\pi\eta')^{-1/2}e^{-a^2\eta(x)}\left(1+O\left(\frac{1}{a^2}\right)\right), \quad V_2 = \frac{1}{a}\left(\frac{2\eta'}{\pi}\right)^{-1/2}e^{a^2\eta(x)}\left(1+O\left(\frac{1}{a^2}\right)\right) \tag{4.50}$$

[$\eta(x)$ is defined by Eq. (4.8)] and

$$W_2 = \left(\frac{d\xi}{dx}\right)^{-1/2} W_{\varkappa,\nu/2}(a^2\xi(x))\left(1+O\left(\frac{1}{a^2}\right)\right) = (\eta')^{-\frac{1}{2}}\varkappa^{\varkappa}\exp\left(-\varkappa+a^2\eta(\lambda)-a^2\eta(x)\right)\left(1+O\left(\min\left(\frac{1}{\varkappa}, \frac{\varkappa^2+1}{a^2}\right)\right)\right), \tag{4.51}$$

$$W_3 = \left(\frac{d\xi}{dx}\right)^{-1/2} W_{-\varkappa,\nu/2}(a^2\xi(x)e^{-i\pi})\left(1+O(1/a^2)\right) = (\eta')^{-1/2}\varkappa^{-\varkappa}\exp\left(\varkappa-a^2\eta(\lambda)+a^2\eta(x)+i\varkappa\pi\right)\left(1+O\left(\min\left(\frac{1}{\varkappa}, \frac{\varkappa^2+1}{a^2}\right)\right)\right). \tag{4.52}$$

Equations (4.50) ensue directly from the well-known asymptotic expansions of Bessel functions; we now prove (4.51) and (4.52). If $\varkappa$ is sufficiently large, we have by virtue of (3.64), (3.27), and (3.58)

$$W_{\varkappa,\nu/2}(a^2\xi) = \left(\frac{\varkappa}{e}\right)^{\varkappa}\left(1-\frac{4\varkappa}{a^2\xi}\right)^{-1/4}\exp\left(-2\varkappa\int_1^{\frac{\xi}{4\gamma^2}}\sqrt{1-\frac{1}{s}}\,ds\right)\left(1+O\left(\frac{1}{\varkappa}+\frac{1}{a^2}\right)\right). \tag{4.53}$$

According to Eq. (4.3), however,

$$\int_1^{\xi/4\gamma^2}\sqrt{1-\frac{1}{s}}\,ds = \frac{1}{2\gamma^2}\int_\lambda^x\sqrt{\frac{s^2-\lambda^2}{1-s^2}}\,ds = \frac{1}{2\gamma^2}\left(-\eta(\lambda)+\eta(x)\right)$$

and $\left(\frac{d\xi}{dx}\right)^2\left(1-\frac{4\varkappa}{a^2\xi}\right) = \left(\frac{d\eta}{dx}\right)^2$; substituting these equations into (4.53), we arrive at (4.51); the asymptotic representation for W_2 is obtained (for large $\varkappa$) in exactly the same way. If, on the other hand, $\varkappa$ is small, we have by (3.55)

$$W_{\varkappa,\nu/2}(a^2\xi) = (a^2\xi)^{\varkappa}e^{-\frac{1}{2}a^2\xi}\left(1+O\left(\frac{1+\varkappa^2}{a^2}\right)\right).$$

In place of the right-hand side we now write

$$\left(1-\frac{4\gamma^2}{\xi}\right)^{-1/4}\exp\left(-2\varkappa\int_1^{\xi/4\gamma^2}\sqrt{1-\frac{1}{s}}\,ds+\varkappa\ln\varkappa-\varkappa\right)\left(1+O\left(\frac{1+\varkappa^2}{a^2}\right)\right),$$

since

$$2\varkappa\int_1^{\xi/4\gamma^2}\sqrt{1-\frac{1}{s}}\,ds=\frac{\varkappa\xi}{2\gamma^2}-\varkappa\ln\frac{\xi}{\gamma^2}-\varkappa+O\left(\frac{\gamma^2\varkappa}{\xi}\right)=\frac{a^2}{2}\xi-\varkappa\ln\frac{a^2\xi}{\varkappa}-\varkappa+O\left(\frac{\varkappa^2}{a^2}\right).$$

Consequently, (4.51) holds either for large or for small $\varkappa$, and Eqs. (4.52) are proved analogously.

Inserting the asymptotic equations into (4.48), we obtain, correct to the factor $\left(\min\left(\frac{1+\varkappa^2}{a^2},\frac{1}{\varkappa}\right)\right)$,

$$\int_{x_1}^{x_2}\frac{dx}{W_3^2}=\varkappa^{2\varkappa}e^{-3\varkappa+a^2\eta(\lambda)-2i\pi\varkappa}\int_{x_1}^{x_2}e^{2a^2\eta(x)}\eta'(x)\,dx. \tag{4.49}$$

The transition from (4.48) to (4.49) is now obvious, and in view of (4.50)-(4.52) we obtain for the right-hand side of (4.49)

$$\mathfrak{W}[V_1,W_2]=\left(\frac{2}{\pi}\right)^{1/2}a\exp\left(a^2\int_\lambda^1\sqrt{\frac{s^2-\lambda^2}{1-s^2}}\,ds-\varkappa\ln\varkappa+\varkappa+i\pi\varkappa\right)\left(1+O\left(\min\left(\frac{\varkappa^2+1}{a^2},\frac{1}{\varkappa}\right)\right)\right). \tag{4.54}$$

In exactly the same way we find

$$\mathfrak{W}[V_2',W_2]=-\sqrt{2\pi}\,a\exp\left(-a^2\int_\lambda^1\sqrt{\frac{s^2-\lambda^2}{1-s^2}}\,ds+\varkappa\ln\varkappa-\varkappa\right)\left(1+O\left(\min\left(\frac{\varkappa^2+1}{a^2},\frac{1}{\varkappa}\right)\right)\right). \tag{4.55}$$

Then, substituting the asymptotic equations (4.50)-(4.52) into the identity

$$\mathfrak{W}[V_1,W_2]\int_{x_1}^{x_2}\frac{dx}{V_1W_2}=\ln\left(\frac{W_2(x_2)V_1(x_1)}{W_2(x_1)V_1(x_2)}\right), \tag{4.56}$$

we obtain

$$\mathfrak{W}[V_1,W_2]=\varkappa^{\varkappa}e^{-\varkappa+a^2\eta(\lambda)-2a^2\eta(x_2)}\,O\left(\min\left(\frac{\varkappa^2}{a},\frac{a}{\varkappa}\right)\right), \tag{4.57}$$

and, in identical fashion,

$$\mathfrak{W}[V_2,W_3]=\varkappa^{-\varkappa}e^{\varkappa-a^2\eta(\lambda)+2a^2\eta(x_1)}\,O\left(\min\left(\frac{\varkappa^2}{a},\frac{a}{\varkappa}\right)\right). \tag{4.58}$$

Substituting the resulting values of the eight Wronskians into (4.36)-(4.38), we obtain the asymptotic behavior of the finite-at-zero solution on the entire interval [0, 1). If now we are to obtain the asymptotic equations for the eigenfunctions, we need merely choose values of λ so that the coefficient in front of $V_2(x)$ in Eq. (4.38) will vanish, and then normalize the resulting solution.

4.7. Equations for the Eigenvalues. The Wronskians of the functions V_1, V_2 and W_1, W_3 on the left-hand side of Eq. (4.39) have nonzero values for all investigated values of the parameters. The equation $A_1=0$ is therefore equivalent to the equation

$$\mathfrak{W}[W_1,W_2]=\frac{\mathfrak{W}[W_1,W_3]\,\mathfrak{W}[W_2,V_1]}{\mathfrak{W}[W_3,V_1]}$$

or, if we substitute the values of the Wronskians into the right-hand side,

$$\mathfrak{J}[W_1, W_2] = \frac{\exp(2\varkappa \ln\varkappa - 2\varkappa + 2a^2\varphi(\lambda) - 2a^2\varphi(x_2))}{\Gamma(\varkappa + \nu/2 + 1/2)} O\left(\min\left(\varkappa^2, \frac{a^2}{\varkappa}\right)\right). \tag{4.59}$$

Making use of Eq. (4.46) and the inequality $\Gamma(x+1/2) \leq c\left(\frac{x}{e}\right)^x$ which holds for $x \geq 0$, we at first obtain from Eq. (4.59)

$$-\frac{\pi \mathfrak{J}[W_1, W_2]}{a^2\Gamma(\nu+1)\Gamma(\varkappa - \nu/2 + 1/2)} = \cos\pi(\varkappa - \nu/2) + O\left(\frac{1}{a^2}\right) = \exp\left(-2a^2\int_\lambda^{x_2}\sqrt{\frac{s^2-\lambda^2}{1-s^2}}\,ds\right) O\left(\min\left(\frac{\varkappa^2}{a^2}, \frac{1}{\varkappa}\right)\right). \tag{4.60}$$

This result implies that the quantity $\varkappa - \nu/2 - 1/2$ differs from an integer only by the term $O(1/a^2)$. This integer, say n, must be chosen so that the corresponding eigenfunction will have exactly n zeros in the interval $(0, a)$. In order to accomplish this we note that the stated problem for the Whittaker equation is close to the boundary-value problem on the semiaxis $[0, +\infty)$ for sufficiently large a. In this case, as is immediately implied by (3.53) [where the coefficient in front of $W_{-\varkappa,\mu}(\varkappa e^{i\pi})$ vanishes on the semiaxis for an eigenfunction of the problem], the equations for the eigenvalues have the form $\varkappa - \nu/2 - 1/2 = n$, $n = 0, 1, 2, \ldots$. Thus, the choice of the constant component in n is uniquely determined, and the following is a consequence of (4.60).

Lemma 11. The eigenvalues λ_n, $n = 0, 1, 2, \ldots$, of the boundary-value problem for Eq. (4.2) satisfy the asymptotic equation

$$\varkappa_n - \nu/2 - 1/2 + O\left(\frac{1}{a^2}\right) \equiv \frac{a^2}{\pi}\int_0^{\lambda_n}\sqrt{\frac{\lambda_n^2 - s^2}{1-s^2}}\,ds - \frac{\nu}{2} - \frac{1}{2} + O\left(\frac{1}{a^2}\right) = n. \tag{4.61}$$

4.8. Asymptotic Equations for the First Eigenfunctions.

We begin by noting that Eqs. (4.59) and (4.47) imply that the coefficient in front of $W_3(x)$ in (4.37) is estimated by the quantity

$$\exp\left((\varkappa - \nu/2)\ln\varkappa - \varkappa - 2a^2\int_\lambda^{x_2}\sqrt{\frac{s^2-\lambda^2}{1-s^2}}\,ds\right) O\left(\min\left(\frac{\varkappa^2}{a^2}, \frac{1}{\varkappa}\right)\right).$$

The coefficient in front of $W_2(x)$ in (4.37) is equal, by virtue of (4.47) and (4.61), to

$$\frac{e^{-i\pi(\varkappa - \nu/2 - 1/2)}\Gamma(\nu+1)}{\Gamma(\varkappa + \nu/2 + 1/2)}\left(1 + O\left(\frac{1}{a^2}\right)\right) = \frac{(-1)^n\Gamma(\nu+1)}{\Gamma(n+\nu+1)}\left(1 + O\left(\frac{\ln(1+n)}{a^2}\right)\right), \tag{4.62}$$

and it is quickly verified by means of the asymptotic equations (4.51) and (4.52) that the right-hand side of (4.37) can be written as follows for $x \leq x_3$ [x_3 is any number from the interval $(1-2\delta, 1-\delta)$]:

$$\frac{(-1)^n\Gamma(\nu+1)}{\Gamma(n+\nu+1)}\left(\frac{d\xi}{dx}\right)^{-\frac{1}{2}} W_{\varkappa_n, \nu/2}(a^2\xi(x, \lambda_n))\left(1 + O\left(\frac{\ln(n+1)}{a^2}\right)\right). \tag{4.63}$$

Then $A_1 = 0$ in Eq. (4.38), and from (4.36)-(4.37) and (4.54)-(4.55) we obtain for the coefficient A_2

$$A_2 = \frac{(-1)^n\sqrt{2\pi}\,\Gamma(\nu+1)}{\Gamma(n+\nu+1)} a\exp\left(-a^2\int_{\lambda_n}^1\sqrt{\frac{s^2-\lambda_n^2}{1-s^2}}\,ds + \varkappa_n\ln\varkappa_n - \varkappa_n\right)\left(1 + O\left(\min\left(\frac{1}{n}, \frac{(1+n^2)\ln(1+n)}{a^2}\right)\right)\right). \tag{4.64}$$

We now compute the normalized integral for the eigenfunction. We normalize the eigenfunctions so that $\int_0^a \psi_n^2\,dx = 1$; consequently, an eigenfunction of Eq. (4.2) must be normalized by the condition

$$a\int_0^1 \frac{w^2(x)}{1-x^2}\,dx = 1. \tag{4.65}$$

For the function $w(x)$ defined by Eqs. (4.36)-(4.38) [with coefficients from (4.62)-(4.64)] this normalized integral is equal to

$$a\left(1+O\left(\frac{1}{a^2}\right)\right)\int_0^\lambda \frac{M^2_{\varkappa,\nu/2}(a^2\xi(x,\lambda_n))dx}{(1-x^2)\xi'}+a\left(1+O\left(\frac{1}{a^2}\right)\right)\int_\lambda^{x_2-\delta}\frac{\Gamma^2(\nu+1)}{\Gamma^2(n+\nu+1)}\frac{W^2_{\varkappa,\nu/2}(a^2\xi(x,\lambda_n))}{(1-x^2)\xi'}dx+a\,A_2^2\int_{x_2-\delta}^{1}\frac{\eta(x)}{(x^2-1)\eta'(x)}\,I^2(-a^2\eta(x))dx,$$

(4.66)

where the number x_2 is chosen from the interval $\frac{1}{\sqrt{2}}+2\delta\le x_2\le 1-4\delta$. For the computation of these integrals we consider the integral

$$\Phi_1=\int_0^\lambda\frac{\lambda^2-x^2}{1-x^2}\left(\frac{d\xi}{dx}\right)^{-1}M^2_{\varkappa,\nu/2}(a^2\xi(x))dx.$$

According to Eq. (4.3) we have for the function $\xi(x,\lambda)$

$$4\,\frac{\lambda^2-x^2}{1-x^2}\left(\frac{d\xi}{dx}\right)^{-2}=\frac{4\gamma^2-\xi}{\xi}.$$

The integral Φ_1 is therefore equal to

$$\frac{1}{4}\int_0^{4\gamma^2}\left(\frac{4\gamma^2}{\xi}-1\right)M^2_{\varkappa,\nu/2}(a^2\xi)d\xi.$$

Differentiating this expression for Φ_1 with respect to λ, we obtain (recalling that $\gamma^2=\frac{1}{\pi}\int_0^\lambda\sqrt{\frac{\lambda^2-s^2}{1-s^2}}\,ds$)

$$\int_0^\lambda\frac{M^2_{\varkappa,\nu/2}(a^2\xi(x,\lambda))}{1-x^2}\frac{dx}{\xi'}=\frac{\gamma}{\lambda}\frac{\partial\gamma}{\partial\lambda}\int_0^{4\gamma^2}M^2_{\varkappa,\nu/2}(a^2\xi)\frac{d\xi}{\xi}. \tag{4.67}$$

Analogously,

$$\int_\lambda^{x_2-\delta}\frac{W^2_{\varkappa,\nu/2}(a^2\xi(x,\lambda))}{1-x^2}\frac{dx}{\xi'}=\frac{\gamma}{\lambda}\frac{\partial\gamma}{\partial\lambda}\int_{4\gamma^2}^{\xi(x_2-\delta,\lambda)}W^2_{\varkappa,\nu/2}(a^2\xi)\frac{d\xi}{\xi}.$$

Furthermore, if we write $n+\nu/2+1/2$ in place of $\varkappa_n$ in these integrals [i.e., reject the remaining terms in Eq. (4.61)], we readily verify that the values of the integrals will only be changed by a factor of the form $1+O\left(\frac{\ln(1+n)}{a^2}\right)$. Consequently, it suffices to compute the integral

$$\int_0^{\xi(x_2-\delta,\lambda_n)}M^2_{n+\nu/2+1/2,\,\nu/2}(a^2\xi)\frac{d\xi}{\xi}=\left(\int_0^\infty-\int_{a^2\xi(x_2-\delta,\lambda_n)}^\infty\right)M^2_{\varkappa_n,\nu/2}(\xi)\frac{d\xi}{\xi}. \tag{4.68}$$

The second term here is estimated, due to (4.52), by the quantity $O(e^{-\delta a^2})$ and the first integral has been computed in [28], p. 116; it is equal to $\frac{n!\,\Gamma^2(\nu+1)}{\Gamma(n+\nu+1)}$*. Finally, the third term in (4.66), as is quickly checked, is estimated by the quantity $O(e^{-\delta a^2})$; the normalized integral is therefore equal to

$$\mathfrak{D}_n^2=\frac{a}{2\pi}\int_0^1\frac{ds}{\sqrt{(1-s^2)(1-\lambda_n^2s^2)}}\,\frac{n!\,\Gamma^2(\nu+1)}{\Gamma(n+\nu+1)}\left(1+O\left(\frac{\ln(n+1)}{a^2}\right)\right). \tag{4.69}$$

Now by simply combining the foregoing estimates we deduce the following inequalities for the normalized eigenfunctions of the boundary-value problem (4.1).

---

*It is important to realize that the normalization of the Whittaker functions in [28] does not concur with the normalization adopted in [6] and in the present article.

Theorem 18. Let $\psi_n(x,a)$ be the nth [in order of increasing number of zeros in the interval $(0,a)$] eigenfunction of the boundary-value problem

$$(x^2-a^2)\frac{d^2\psi}{dx^2}+2x\frac{d\psi}{dx}+a^2\left(x^2+\frac{\nu^2-1/4}{x^2}\right)\psi=a^4\lambda^2\psi, \quad |\psi(0)|,\ |\psi(a)|<\infty .$$

If $\lambda \leq 1/\sqrt{2}$ and a is sufficiently large, then for

$$\mathfrak{D}_n\psi_n(ax,a)=\left((1-x^2)\frac{d\xi}{dx}\right)^{-1/2}\left(M_{n+\nu/2+1/2,\,\nu/2}(a^2\xi(x,\lambda_n))+O\left(\frac{\min\left(a^\nu\xi^{\nu/2+1},\ \frac{1}{a}\sqrt{\frac{\xi}{1+n}}\right)}{(1+n)^{-1/6}+(a^2(1+n)\xi)^{-\nu/2}+\left|1-\frac{4n+2\nu+2}{a^2\xi}\right|^{1/4}}\right)\right), \tag{4.70}$$

for $\lambda_n \leq x \leq 1-\delta$

$$\mathfrak{D}_n\psi_n(ax,a)=\frac{(-1)^n\Gamma(\nu+1)}{\Gamma(n+\nu+1)}\left((1-x^2)\frac{d\xi}{dx}\right)^{-\frac{1}{2}}\left(1+O\left(\frac{\ln(1+n)}{a^2}\right)\right)W_{n+\nu/2+1/2,\,\nu/2}(a^2\xi(x,\lambda_n)), \tag{4.71}$$

for $\frac{1}{\sqrt{2}} \leq x \leq 1$

$$\mathfrak{D}_n\psi_n(ax,a)\left(1+O\left(\min\left(\frac{1+n^2}{a^2},\frac{1}{n}\right)\right)\right)=\frac{(2\pi)^{1/2}(-1)^n\Gamma(\nu+1)}{\Gamma(n+\nu+1)}\cdot a\exp\left(-a^2\int_{\lambda_n}^{1}\sqrt{\frac{s^2-\lambda_n^2}{1-s^2}}\,ds+\varkappa_n\ln\varkappa_n-\varkappa_n\right)\left(\frac{-\eta(x)}{(1-x^2)\eta'(x)}\right)^{1/2}I_0(-a^2\eta(x,\lambda_n)), \tag{4.72}$$

and for

$$\mathfrak{D}_n\psi_n(ax,a)\left(1+O\left(\min\left(\frac{1+n^2}{a^2},\frac{1}{n}\right)\right)\right)=\frac{(-1)^n(2\pi)^{1/2}a\Gamma(\nu+1)}{\Gamma(n+\nu+1)}\exp\left(-a^2\int_{\lambda_n}^{1}\sqrt{\frac{s^2-\lambda_n^2}{1-s^2}}\,ds+\varkappa_n\ln\varkappa_n-\varkappa_n\right)\left(\frac{\eta}{(x^2-1)\eta'}\right)^{1/2}\left(J_0(a^2\eta)+O\left(\frac{\eta^2}{(1+a\sqrt{\eta})(1+a^2\eta(1+\eta))}\right)\right), \tag{4.73}$$

where the numbers λ_n are determined from Eqs. (4.61), $\mathfrak{D}_n$ is determined from (4.69), and the functions $\xi(x,\lambda)$ and $\eta(x,\lambda)$ are given by the respective equations (4.3) and (4.8).

§ 5. Passage to the Limit as $a \to \infty$

5.1. Limit Functions. In this section we go to the limit as $a \to +\infty$ in the integral equation

$$\mu\,\psi(x)=\int_0^a\sqrt{x\xi}\,J_\nu(x\xi)\,\psi(\xi)\,d\xi, \quad \nu>\tfrac{1}{2}. \tag{5.1}$$

In § 4 we obtained asymptotic equations for the eigenfunctions $\psi_n(x,a)$ of this equation under the condition $n \leq A_0 a^2$, $A_0=\frac{1}{2a}\int_0^1\sqrt{\frac{1-s^2}{1-s^{1/2}}}\,ds$; we now prove that the following is a consequence of these equations (Theorem 18, Sec. 4.8).

Theorem 19. For every fixed n we have the following, uniformly on x in every finite interval:

$$\lim_{a\to\infty}\psi_n(x,a)=\left(\frac{\Gamma(n+\nu+1)}{\Gamma^2(\nu+1)n!}\right)^{1/2}\sqrt{\frac{2}{x}}\,M_{n+\nu/2+1/2,\,\nu/2}(x^2), \tag{5.2}$$

where $M_{\varkappa,\nu}(z)$ is the Whittaker function

$$M_{n+\nu/2+1/2,\,\nu/2}(z)=z^{\frac{\nu+1}{2}}e^{-\frac{1}{2}z}\sum_{k=0}^{n}\frac{n!\,\Gamma(\nu+1)(-z)^k}{k!\,(n-k)!\,\Gamma(k+\nu+1)} . \tag{5.3}$$

We denote the right-hand side of (5.2) for brevity by $\varphi_n(x)$, $n=0,1,\ldots$. These functions are significant in connection with the fact that they play the same role in the theory of the νth-order Hankel transform as do the Hermite functions in the theory of the Fourier transform (see [3], pp. 102-110). More specifically, the following propositions are true.

Theorem 20. The function $\varphi_n(x)$ forms a normal orthogonal system on the semiaxis $[0,+\infty)$, i.e.,

$$\int_0^\infty \varphi_n(x)\varphi_m(x)dx = \begin{cases} 0, & n \neq m; \\ 1, & n = m. \end{cases}$$

Theorem 21. The νth-order Hankel transform of the function $\varphi_n(x)$ is equal to $(-1)^n\varphi_n(x)$, i.e.,

$$\int_0^\infty \sqrt{x\xi}\, J_\nu(x\xi)\varphi_n(\xi)d\xi = (-1)^n\varphi_n(x). \tag{5.4}$$

Theorem 22. Let $f(x)$ be an arbitrary function from $\mathcal{L}_2(0,\infty)$. Let us put $b_n = \int b_n = \int_0^\infty f(x)\varphi_n(x)\,dx$. Then

$$\lim_{N\to\infty}\int_0^\infty \left|f(x)-\sum_{n=0}^N b_n\varphi_n(x)\right|^2 dx=0. \tag{5.5}$$

We shall omit the proofs of Theorems 20 and 22, as they differ only in notation from the corresponding propositions of the theory of Laguerre polynomials. Thus, the Laguerre polynomial $L_n^\alpha(x)$ is equal to ([29], pp. 116-119)

$$L_n^\alpha(x)=\frac{x^{-\alpha}e^x}{n!}\frac{d^n}{dx^n}\left(x^{n+\alpha}e^{-x}\right)=\sum_{k=0}^n\frac{\Gamma(n+\alpha+1)}{(n-k)!\,\Gamma(k+\alpha+1)}\frac{(-x)^k}{k!}, \tag{5.6}$$

and comparison with (4.2) reveals that

$$\varphi_n(x)=\left(\frac{2n!}{\Gamma(n+\nu+1)}\right)^{1/2}x^{\nu+1/2}e^{-1/2x^2}L_n^\nu(x^2). \tag{5.7}$$

The orthogonality and completeness properties of the functions $\varphi_n(x)$ with unit weight therefore ensue from the corresponding properties of the Laguerre polynomials with weight $x^\nu e^{-x}$.

Equation (5.4) also follows from the familiar properties of the Laguerre polynomials. We know that the generating function for the Laguerre polynomials has the form ([29], p. 111)

$$\sum_{n=0}^\infty z^n L_n^\alpha(x)=(1-z)^{-\alpha-1}e^{\frac{zx}{z-1}} \qquad |z|<1. \tag{5.8}$$

Therefore, by virtue of (5.7)

$$\sum_{n=0}^\infty\left(\frac{\Gamma(n+\nu+1)}{n!}\right)^{1/2}z^n\varphi_n(x)=2^{1/2}(1-z)^{-\nu-1}x^{\nu+1/2}e^{-\frac{x^2}{2}\frac{1+z}{1-z}}. \tag{5.9}$$

We multiply both sides of this equation by $\sqrt{x\xi}\,J_\nu(x\xi)$ and integrate term by term between the limits from 0 to ∞; according to the following well-known equation from the theory of Bessel functions ([30], p. 60, function (24)):

$$\int_0^\infty J_\nu(x\xi)e^{-\frac{1}{2}\gamma x^2}x^{\nu+1}dx=\xi^\nu\gamma^{-\nu-1}e^{-\xi^2/2\gamma}, \quad \operatorname{Re}\gamma>0; \tag{5.10}$$

we obtain upon integration of (5.9)

$$\sum_{n=0}^\infty\left(\frac{\Gamma(n+\nu+1)}{n!}\right)^{1/2}z^n\int_0^\infty\sqrt{x\xi}\,J_\nu(x\xi)\varphi_n(x)dx=2^{1/2}(1+z)^{-\nu-1}\xi^{\nu+1/2}e^{-\frac{\xi^2}{2}\frac{1-z}{1+z}}. \tag{5.11}$$

Replacing z by $-z$ in the latter and comparing the resulting equation with (5.9), we at once deduce (5.4).

Next we prove (5.2). We recall that one of the propositions of the cited Theorem 18 states that for $x \leq 1/2$ and $n \leq A_0 a^2$

$$\mathcal{D}_n \psi_n(ax, a) = \begin{cases} \left((1-x^2)\dfrac{d\xi(x,\lambda_n)}{dx}\right)^{-1/2}\left(M_{n+\nu/2+1/2,\,\nu/2}(a^2\xi(x,\lambda_n)) + O\left(\dfrac{\sqrt{\xi(x,\lambda_n)}}{a(1+n)^{1/3}}\right)\right), & 0 \leq x \leq \lambda_n; \quad (5.12) \\[2em] \dfrac{(-1)^n\Gamma(\nu+1)}{\Gamma(n+\nu+1)}\left((1-x^2)\dfrac{d\xi(x,\lambda_n)}{dx}\right)^{-1/2} W_{n+\nu/2+1/2,\,\nu/2}(a^2\xi(x,\lambda_n))\left(1+O\left(\dfrac{\ln(1+n)}{a^2}\right)\right), & \lambda_n \leq x \leq \frac{1}{2}, \quad (5.13) \end{cases}$$

in which the conventional notation for the Whittaker functions is used, λ_n and $\mathcal{D}_n$ are determined from the respective equations (4.61) and (4.69), and the function $\xi(x, \lambda_n)$ satisfies the equation

$$\left(\frac{d\xi}{dx}\right)^2\left(\frac{4\gamma_n^2}{\xi} - 1\right) = 4\,\frac{\lambda_n^2 - x^2}{1-x^2}, \quad \xi(0) = 0, \quad \xi(\lambda_n) = 4\gamma_n^2. \tag{5.14}$$

It quickly follows from (4.61) that for fixed n and $a \to \infty$

$$\frac{1}{4}a^2\lambda_n^2 = n + \frac{\nu}{2} + \frac{1}{2} + O\left(\frac{1+n^2}{a^2}\right). \tag{5.15}$$

Consequently, the function $\tilde{\xi}(x) = a\sqrt{\xi\left(\frac{x}{a}, \lambda_n\right)}$ satisfies the equation

$$\left(4n + 2\nu + 2 + O\left(\frac{1}{a^2}\right) - \tilde{\xi}^2\right)\left(\frac{d\tilde{\xi}}{dx}\right)^2 = 4n + 2\nu + 2 + O\left(\frac{1}{a^2}\right) - x^2. \tag{5.16}$$

Thus, for every fixed x we have (uniformly with respect to x in every finite interval) $\tilde{\xi} \to x$ as $a \to +\infty$, i.e.,

$$\lim_{a\to\infty} a^2\xi\left(\frac{x}{a}, \lambda_n\right) = x^2. \tag{5.17}$$

The remainder term in (5.12) is therefore estimated by the quantity $O\left(\frac{x}{a^2(1+n)^{1/3}}\right)$; also,

$$(1-t^2)\frac{d\xi(t,\lambda_n)}{dt}\bigg|_{t=\frac{x}{a}} = \frac{2x}{a}\left(1 + O\left(\frac{1}{a^2}\right)\right), \tag{5.18}$$

and

$$\mathcal{D}_n^2 = \frac{a}{4}\,\frac{n!\,\Gamma(\nu+1)}{\Gamma(n+\nu+1)}\left(1 + O\left(\frac{1}{a^2}\right)\right). \tag{5.19}$$

Substituting the resulting estimates into (5.12) and (5.13) and replacing x by x/a, we immediately deduce the statement of the theorem.

§ 6. Asymptotic Equations for the First Eigenvalues of the Integral Equation

We have already noted in Sec. 2.6 the existence of $\lim\limits_{a\to\infty} \mu_n$ for every fixed n. A comparison of Eqs. (5.2) and (5.4) shows that this limit is equal to $(-1)^n$. We now estimate the difference $\mu_n - (-1)^n$ for large a and $n \leq A_0 a^2$.

Theorem 23. Let a be sufficiently large, and let $n \leq A_0 a^2$, $A_0 = \frac{1}{2\pi}\int_0^1 \sqrt{\frac{1-s^2}{1-s^2/2}}\,ds$; then the following equation holds for the n eigenvalue of the integral equation (5.1):

$$\mu_n = (-1)^n + \frac{2\pi(-1)^{n+1}}{n!\,\Gamma(n+\nu+1)} \exp\left(2\varkappa_n \ln \varkappa_n - 2\varkappa_n - 2\int_{\lambda_n}^1 \sqrt{\frac{s^2-\lambda_n^2}{1-s^2}}\,ds\,a^2\right)\left(1+O\left(\min\left(\frac{1+n^2}{a^2},\frac{1}{n}\right)\right)\right), \tag{6.1}$$

where $\varkappa_n = n + \nu/2 + 1/2$. In particular, for $a^{2/3} \leq n \leq A_0 a^2$

$$\mu_n = (-1)^n\left(1 - \exp\left(-2a^2\int_{\lambda_n}^1 \sqrt{\frac{s^2-\lambda_n^2}{1-s^2}}\,ds\right)\left(1+O\left(\frac{1}{n}\right)\right)\right). \tag{6.2}$$

For the proof we use an equation from Sec. 2.6 and Theorem 18 from Sec. 4.8. We have

$$\frac{d\mu_n}{da} = \mu_n \psi_n^2(a,a)$$

and Eq. (4.72) gives the following for $x = 1$:

$$\psi_n^2(a,a) = aB_n^2\left(\int_0^1 \frac{ds}{\sqrt{(1-s^2)(1-\lambda_n^2 s^2)}}\right)^{-1} \exp\left(-2a^2\int_{\lambda_n}^1 \sqrt{\frac{s^2-\lambda_n^2}{1-s^2}}\,ds\right)\left(1+O\left(\min\left(\frac{1+n^2}{a^2},\frac{1}{n}\right)\right)\right), \tag{6.3}$$

where

$$B_n^2 = 4\pi^2 \frac{e^{2\varkappa_n \ln \varkappa_n - 2\varkappa_n}}{n!\,\Gamma(n+\nu+1)}. \tag{6.4}$$

Let us compute the integral $\Phi = \int_a^\infty \psi_n^2(x,x)\,dx$; then Eq. (6.1) will follow directly from the equation

$$\mu_n(a) = \mu_n(\infty)e^{-\Phi} = (-1)^n e^{-\Phi}. \tag{6.5}$$

We note at the outset that λ_n may be interpreted in (6.3) as the solution of Eq. (4.61) without remainder terms. Thus, each of the λ_n changes at most by $O(1/a^4)$ when the remainder term $O(1/a^2)$ is discarded in (4.61). Therefore, the integral in the exponent of (6.3) changes by $O(1/a^2)$ in this case, i.e., the entire right-hand side of (6.3) is multiplied only by a factor of the form $1+O\left(\frac{1}{a^2}\right)$.

For the calculations we use the Jacobi elliptic functions ([6], Chap. 22). We make a change of the variable of integration, $s = \mathrm{dn}(z, \sqrt{1-\lambda_n^2})$, in the integral contained in the exponent of (6.3); recognizing that as z varies from zero to $K' = \int_0^1 (1-s^2)^{-1/2}(1-(1-\lambda_n^2)s^2)^{-1/2}\,ds$ the function $\mathrm{dn}(z,\sqrt{1-\lambda_n^2})$ varies from unity to λ_n, we obtain

$$\int_{\lambda_n}^1 \left(\frac{s^2-\lambda_n^2}{1-s^2}\right)^{1/2} ds = (1-\lambda_n^2)\int_0^{K'} \mathrm{cn}^2(z,\sqrt{1-\lambda_n^2})\,dz = -\lambda_n^2 K' + \int_0^{K'} \mathrm{dn}^2(z,\sqrt{1-\lambda_n^2})\,dz \tag{6.6}$$

(here we use the algebraic relations between Jacobi elliptic functions and the differentiation rule; see [6], pp. 376-377). We denote the last term in (6.6) by E' and let K and E be the same functions of λ_n as K' and E' are of $\sqrt{1-\lambda_n^2}$. In this notation the integral that we need to compute assumes the form

$$\int_a^\infty \psi_n^2(x,\lambda)dx = B_n^2\left(1+O\left(\min\left(\frac{1+n^2}{a^2},\frac{1}{n}\right)\right)\right)\int_a^\infty \exp\left(-2x^2E' + 2x^2\lambda_n^2K'\right)\frac{x\,dx}{K}, \tag{6.7}$$

where the function $\lambda_n(x)$ satisfies the equation

$$\begin{aligned} n+\nu/2+1/2 &= \frac{x^2}{\pi}\int_0^{\lambda_n}\sqrt{\frac{\lambda_n^2-s^2}{1-s^2}}\,ds \\ &= \frac{x^2\lambda_n^2}{\pi}\int_0^K \mathrm{cn}^2(z,\lambda_n)dz \\ &= \frac{x^2}{\pi}\left(-(1-\lambda_n^2)K+\int_0^K \mathrm{dn}^2(z,\lambda_n)dz\right) \end{aligned} \tag{6.8}$$

[here we have incorporated the change of integration variable $s=\lambda_n \mathrm{sn}(z,\lambda_n)$]. It follows from (6.8) that the derivative $\frac{\partial \lambda_n}{\partial x}$ satisfies the equation

$$-2(1-\lambda_n^2)K + 2E + x\lambda_n K\frac{\partial \lambda_n}{\partial x} = 0. \tag{6.9}$$

Also,

$$\frac{\partial}{\partial x}\left(x^2\int_{\lambda_n}^1\sqrt{\frac{s^2-\lambda_n^2}{1-s^2}}\,ds\right) = 2x(-\lambda_n^2K'+E') - x^2\lambda_n\frac{\partial\lambda_n}{\partial x}\int_{\lambda_n}^1\frac{ds}{\sqrt{(s^2-\lambda_n^2)(1-s^2)}} = 2x(-\lambda_n^2K'+E') - x^2\lambda_n\frac{\partial\lambda_n}{\partial x}K'. \tag{6.10}$$

Inserting the value for $\frac{\partial\lambda_n}{\partial x}$, we obtain

$$\frac{\partial}{\partial x}\left(x^2\int_{\lambda_n}^1\sqrt{\frac{s^2-\lambda_n^2}{1-s^2}}\,ds\right) = \frac{2x}{K}(KE'+EK'-KK'). \tag{6.11}$$

According to the Legendre relation ([6], p. 415) the right-hand side of (6.11) is equal to $\frac{\pi x}{K}$; consequently, the integral on the right-hand side of (6.7) is equal to

$$-\frac{1}{2\pi}\int_a^\infty \frac{d}{dx}\left(e^{-2x^2E'+2x^2\lambda_n^2K'}\right)dx = \frac{1}{2\pi}\exp\left(-2a^2\int_{\lambda_n}^1\sqrt{\frac{s^2-\lambda_n^2}{1-s^2}}\,ds\right). \tag{6.12}$$

Together with (6.4) and (6.5) this equation at once yields (6.1), and the Stirling expansion for large n leads to (6.2).

§ 7. Asymptotic Equations for High-Order Eigenfunctions

7.1. Transformation of the Equations. We recall that the nth eigenfunction of the integral equation (5.1) is the nth eigenfunction of the boundary-value problem

$$(x^2-a^2)\frac{d^2\psi}{dx^2} + 2x\frac{d\psi}{dx} + a^2\left(x^2+\frac{\nu^2-1/4}{x^2}\right)\psi = a^4\lambda^2\psi, \quad |\psi(0)|, |\psi(a)| < \infty. \tag{7.1}$$

Therefore, the function

$$w(x) = \begin{cases} \sqrt{1-x^2}\,\psi(ax), & 0 \le x < 1; \\ \sqrt{x^2-1}\,\psi(ax), & 1 < x < \infty. \end{cases}$$

satisfies the differential equation

$$\frac{d^2w}{dx^2} + \left(a^4\frac{\lambda^2-x^2}{1-x^2} + \frac{1}{(1-x^2)^2} - \frac{\nu^2-1/4}{x^2(1-x^2)}\right)w = 0. \tag{7.2}$$

We have already obtained estimates for the solutions of this equation when $\lambda \leq 1/\sqrt{2}$; we invoke the same method for $\lambda > 1/\sqrt{2}$. Specifically, adopting the appropriate change of variable and unknown function, we transform Eq. (7.2) to the following for small x:

$$\frac{d^2W}{d\eta^2} + \left(a^4 - \frac{\nu^2 - 1/4}{\eta^2}\right)W = f_1(\eta)W, \tag{7.3}$$

and to the following for $1/4 \leq x < \infty$:

$$\frac{d^2V}{d\xi^2} + \left(\frac{a^4}{4} - \frac{\varkappa a^2}{\xi} + \frac{1}{4\xi^2}\right)V = f_2(\xi)V. \tag{7.4}$$

Next we estimate the functions f_1 and f_2 and apply Theorems 10 and 17 from § 3; we infer as a result that the finite-at-zero solution of Eq. (7.2) is close to the function $\left(\frac{d\eta}{dx}\right)^{-1/2} J_\nu\,(a^2\eta)$, and the solution satisfying the boundary condition at the right end differes only slightly from the function $\left(\frac{d\xi}{dx}\right)^{-1/2} M_{i\varkappa,0}$ $(ia^2\xi)$. Then, requiring that these two solutions coincide, we obtain secular equations for the eigenvalues of the boundary-value problem (7.1). This done, we need only normalize the ensuing solution in order to obtain the asymptotic equations for the eigenfunctions.

In order to transform Eq. (7.2) to the form (7.3) we set

$$\eta(x,\lambda) = \int_0^x \sqrt{\frac{\lambda^2 - s^2}{1 - s^2}}\,ds, \qquad 0 \leq x \leq \min\,(\lambda - \delta,\ 1 - \delta) \tag{7.5}$$

and change the variable and unknown function in (7.2), making $w(x) = \left(\frac{d\eta}{dx}\right)^{-1/2} W\,(\eta(x,\lambda))$. A simple calculation shows that $W(\eta)$ satisfies Eq. (7.3), where

$$f_1(\eta) = \left(\frac{1}{2}\{\eta, x\} - \frac{1}{(1-x^2)^2}\right)\left(\frac{dx}{d\eta}\right)^2 + (\nu^2 - 1/4)\left(\frac{1}{x^2(1-x^2)}\left(\frac{dx}{d\eta}\right)^2 - \frac{1}{\eta^2}\right) \tag{7.6}$$

(we denote by $\{\eta, x\}$ the Schwarz derivative of η with respect to x). Clearly, the function $\eta(x,\lambda)$ is monotonically increasing and continuous together with all its derivatives in the interval $0 \leq x \leq \min(\lambda - \delta, 1 - \delta)$ consequently, the inverse function $x = x(\eta, \lambda)$ is endowed with the same properties on the corresponding interval. Therefore, in order to obtain a solution of (7.2) finite at zero we need to choose a solution of (7.3) with the same property.

In order to transform Eq. (7.2) to the form (7.4) we define the function $\xi(x,\lambda)$ by the equation

$$\left(\frac{1}{4} - \frac{\varkappa}{a^2\xi}\right)\left(\frac{d\xi}{dx}\right)^2 = \frac{\lambda^2 - x^2}{1 - x^2}, \quad \frac{1}{4} \leq x < +\infty, \tag{7.7}$$

and the auxiliary conditions

$$\xi(\lambda,\lambda) = \frac{4\varkappa}{a^2}, \qquad \xi(1,\lambda) = 0 \qquad \left(\frac{d\xi(x,\lambda)}{dx} > 0\right), \tag{7.8}$$

which ensure continuity of $\frac{d\xi}{dx}$ at $x = \lambda$ and $x = 1$. The first of these conditions determines the parameter $\varkappa$; integrating (7.7), we find that this condition is met for

$$\varkappa = \begin{cases} -\dfrac{a^2}{\pi}\displaystyle\int_\lambda^1 \sqrt{\frac{s^2 - \lambda^2}{1 - s^2}}\,ds, & \lambda \leq 1; \quad (7.9) \\[2ex] \dfrac{a^2}{\pi}\displaystyle\int_1^\lambda \sqrt{\frac{\lambda^2 - s^2}{s^2 - 1}}\,ds, & \lambda > 1. \quad (7.10) \end{cases}$$

Making the substitution $w(x)=\left(\frac{d\xi}{dx}\right)^{-1/2} V(\xi)$ we obtain Eq. (7.4) for $V(\xi)$ with

$$f_2(\xi)=\left(\tfrac{1}{2}\{\xi,x\}-\frac{1}{(1-x^2)^2}\left(\frac{dx}{d\xi}\right)^2+\frac{1}{4\xi^2}+\frac{\nu^2-1/4}{x^2(1-x^2)}\left(\frac{dx}{d\xi}\right)^2\right). \tag{7.11}$$

The point $\xi=0$ is a regular singular point for Eq. (7.4). For the characteristic equation corresponding to this point, $\frac{1}{2}$ is a multiple root. Hence, one of the solutions of this equation is representable in some neighborhood of the point $\xi=0$ as the product of $\xi^{1/2}$ by a function regular at $\xi=0$, and all other solutions have a singularity of the form $\xi^{1/2}\ln\xi$. We note in addition that when $x\to 1$

$$\xi(x,\lambda)=\frac{(\lambda^2-1)a^2}{4\varkappa}(x^2-1)(1+O(x^2-1)). \tag{7.12}$$

In order, therefore, to obtain a solution of (7.1) regular at $x=1$ we must use a solution of (7.4) that remains finite after division by $\xi^{1/2}$.

7.2. Estimate of $f_1(\eta)$. Direct differentiation of (7.5) yields

$$\{\eta,x\}=-\frac{x^2+2}{2(1-x^2)^2}+\frac{x^2}{(\lambda^2-x^2)(1-x^2)}-\frac{3x^2+2\lambda^2}{2(\lambda^2-x^2)^2}. \tag{7.13}$$

It is clear that for $\lambda\geqslant\frac{1}{\sqrt{2}}$ and $0\leqslant x\leqslant\min(\lambda-\delta,1-\delta)$ the function $\{\eta,x\}$ is bounded (here and elsewhere $\delta>0$- is a fixed sufficiently small number). Then

$$\eta(x,\lambda)=\lambda x+\int_0^\lambda\left(\sqrt{\frac{\lambda^2-s^2}{1-s^2}}-\lambda\right)ds=\lambda x(1+O(x^2)),$$

where the constant in the O-symbol is independent of λ. Consequently,

$$\frac{1}{x^2(1-x^2)}\left(\frac{dx}{d\eta}\right)^2-\frac{1}{\eta^2}=\frac{x^2}{\lambda^2-x^2}-\frac{1+O(x^2)}{\lambda^2x^2}=O\left(\frac{1}{\lambda^2}\right),$$

and the estimate $f_1(\eta)=O(1/\lambda^2)$ holds for the function $f_1(\eta)$ for all $\lambda\geqslant\frac{1}{\sqrt{2}}$ and $0\leqslant x\leqslant\min(\lambda-\delta,1-\delta)$.

7.3. Asymptotic Representation of the Solution Finite at Zero. As a result of the estimate for $f_1(\eta)$ the direct application of Theorem 10 leads to the following.

Lemma 12. *There exists a solution $W(x)$ of Eq. (7.2) such that for $\lambda\geqslant 1/\sqrt{2}$ and $0\leqslant x\leqslant\min(\lambda-\delta,1-\delta)$*

$$W(x)=\left(\frac{d\eta}{dx}\right)^{-1/2}\left(\sqrt{\eta}\,J_\nu(a^2\eta)+O\left(\frac{a^{2\nu}\eta^{\nu+5/2}}{\lambda^2(1+a^2\eta)(1+(a^2\eta)^{\nu+1/2})}\right)\right), \tag{7.14}$$

where the function $\eta=\eta(x,\lambda)$ is determined by Eq. (7.5).

7.4. Estimate of $f_2(\xi)$. In Eq. (7.7) we put $x^2=1+(\lambda^2-1)t$ and $\xi=\frac{4\varkappa}{a^2}\varphi$ for $\lambda\neq 1$. Then the function $\varphi(t,\lambda)$ satisfies the equation

$$\alpha\left(\frac{1}{\varphi}-1\right)\left(\frac{d\varphi}{dt}\right)^2=\frac{1-t}{t(1-(1-\lambda^2)t)}, \tag{7.15}$$

where $\alpha=16\left(\frac{\varkappa}{a^2(\lambda^2-1)}\right)^2$, as well as the auxiliary condition $\varphi(0,\lambda)=0$, $\varphi(1,\lambda)=1$, and the function f_2 is expressed in terms of φ as follows:

$$4\alpha(\lambda^2-1)^2(1+(\lambda^2-1)t)f_2 = \frac{1}{t^2}-\frac{1}{\varphi^2}-\frac{(\lambda^2-1)(\nu^2-\frac{1}{4})}{t(1+(\lambda^2-1)t)}-\frac{(\lambda^2-1)t}{\varphi^2}-\frac{3}{4}\frac{(\lambda^2-1)^2}{1+(\lambda^2-1)t}+2(1+(\lambda^2-1))\{\varphi,t\}. \tag{7.16}$$

This function and Eq. (7.15) for small values of $|\lambda^2-1|$ are completely analogous to their counterparts in Sec. 4.5. It follows at once from the estimates obtained there that, uniformly with respect to λ, $\frac{1}{\sqrt{2}} \leq \lambda \leq \frac{1}{8}$, the function $f_2(\xi)$ is estimated by the quantity $O(\frac{1}{|\xi|})$ for $\frac{1}{4} \leq x \leq \frac{1}{8}$ and by the quantity $O(\frac{1}{\xi^2})$ for $x \to \infty$. It is easily verified that these estimates are valid for all λ not exceeding a fixed constant. If, on the other hand, $\lambda \to \infty$, the following equation applies to the function $f_2(\xi)$:

$$f_2(\xi) = O\left(\frac{\ln\lambda}{\xi(\lambda+|\xi|)}\right). \tag{7.17}$$

For the proof it is sufficient to use the expansions for the function ξ in the neighborhood of the points $x=1$, $x=\lambda$, and $x=+\infty$; we shall omit the corresponding (rather cumbersome) computations from the present discussion.

7.5. Asymptotic Representation of the Solutions in the Interval $\frac{1}{4} \leq x < +\infty$.

If $\varkappa \leq 0$ in Eq. (7.4) (i.e., if $\lambda \leq 1$), the asymptotic behavior of a solution that remains finite after multiplication by $\xi^{-1/2}$ follows at once from Theorem 17 of § 3. The case of nonnegative $\varkappa$ need not be treated separately, because the simultaneous replacement of $\varkappa$ by $-\varkappa$ and of ξ by $-\xi$ (without altering the unperturbed equation) reduces it to the case of $\varkappa \leq 0$. We set

$$\beta(x,\varkappa) = \left(\frac{1}{a|x|^{1/2}}+\frac{1}{1+|\varkappa|^{1/6}}+\left|1-\frac{4\varkappa}{a^2x}\right|^{1/4}\right)^{-1}. \tag{7.18}$$

If $|\varkappa|$ does not exceed a fixed constant, then $\beta(x,\varkappa)=O(\min(a\sqrt{|x|}, 1))$; for the function F involved in the inequality of Theorem 17 (with f replaced by f_2), therefore, we obtain

$$F=O\left(\left|\int_0^\xi \min(a|t|^{1/2}, 1)\min\left(a|t|^{1/2}\ln\left(\frac{1}{a^2|t|}+1\right), 1\right)\min\left(\frac{1}{|t|},\frac{1}{t^2}\right)dt\right|\right)=O\left(\min\left(a^2|\xi|\ln\left(\frac{1}{a^2|\xi|}+1\right), \ln a\right)\right). \tag{7.19}$$

Consequently, $F=O(\ln a)$ uniformly with respect to ξ. For large values of $\varkappa$ we partition the interval $(-\infty,+\infty)$ into subintervals

$$\left(-\infty,\frac{-1}{a^2\varkappa}\right),\left(-\frac{1}{a^2\varkappa},\frac{1}{a^2\varkappa}\right),\left(\frac{1}{a^2\varkappa},\frac{4(1-\delta)\varkappa}{a^2}\right),\left(\frac{4(1-\delta)\varkappa}{a^2},\frac{4(1+\delta)\varkappa}{a^2}\right),\left(\frac{4(1+\delta)\varkappa}{a^2},\infty\right). \tag{7.20}$$

In these subintervals the function $\beta(x,\varkappa)$ is estimated by the respective quantities

$$O(1), O(|x|^{1/2}a), O\left(a^{1/2}\left(\frac{|x|}{\varkappa}\right)^{1/4}\right), O\left(\frac{\varkappa^{1/4}}{\varkappa^{1/12}+|a^2x+4\varkappa|^{1/4}}\right), O(1). \tag{7.21}$$

Estimating the second factor in F analogously and partitioning the interval of integration, we obtain

$$|F(\pm\infty)|=O\left(\frac{(1+|\ln\lambda|)\ln a}{\lambda}\right), \quad \frac{1}{2} \leq \lambda < \infty. \tag{7.22}$$

Observing now that, the coefficients being real, together with $M_{i\varkappa,0}(ia^2\xi)$ the real part of this function* is also a solution of the unperturbed equation (7.4), we arrive at the following result.

---

*Indeed, $\operatorname{Re} M_{i\varkappa,0}(ia^2\xi)=2^{-1/2}e^{-i\pi/4}M_{i\varkappa,0}(ia^2\xi)$ for real $\varkappa$ and $\xi \geq 0$; the same equation, with $e^{-i\pi/4}$ replaced by $e^{i\pi/4}$, also holds for $\xi \leq 0$.

Lemma 13. Given proper normalization, the following equations hold for a solution $V(x)$ of Eq. (7.2) that is left finite after multiplication by $|x^2-1|^{1/2}$, in the interval $1/4 \leq x < \infty$:

$$V(x)=\left(\frac{d\xi}{dx}\right)^{-1/2}\left(\operatorname{Re} M_{i\varkappa,0}(ia^2\xi(x,\lambda))+O\left(\frac{\ln a \ln(1+\lambda)}{a^2\lambda}\right)\beta(\xi,\varkappa)e^{2\varkappa\vartheta\left(\frac{a^2\xi}{4\varkappa}\right)}\right) \tag{7.23}$$

when $\varkappa \geq 0$, and

$$V(x)=\left(\frac{d\xi}{dx}\right)^{-\frac{1}{2}}\left(\operatorname{Re} M_{-i\varkappa,0}(-ia^2\xi(x,\lambda))+O\left(\frac{\ln a}{a^2}\right)\beta(\xi,\varkappa)e^{2\varkappa\vartheta\left(\frac{a^2\xi}{4\varkappa}\right)}\right) \tag{7.24}$$

when $\varkappa < 0$.

We recall that the function $\xi(x,\lambda)$ in these equations is determined by Eq. (7.7), and that $\vartheta(x)=0$ for $x \leq 0$, $\vartheta(x)=\int_0^x\left(\frac{1}{s}-1\right)^{1/2}ds$ for $0 \leq x \leq 1$, and $\vartheta(x)=\pi/2$ for $x \geq 1$.

7.6. Equations for the Eigenvalues. We now "fuse" the solutions from Lemmas 12 and 13. For this it is helpful to consider the behavior of the solutions $V(x)$ and $W(x)$ in the neighborhood of the point $x = 1/4$, since the asymptotic expansion for each of these functions can be used when $x \approx 1/4$.

We begin with the case of large negative $\varkappa$, i.e., large $(1-\lambda^2)a^2$. The following expansion is applicable to the right-hand side of (7.16) for $x \approx 1/4$ and any $\lambda \geq \frac{1}{\sqrt{2}}$:

$$W(x)=\left(\frac{2}{\pi a^2\eta'}\right)^{1/2}\cos\left(a^2\eta(x,\lambda)-\frac{\nu\pi}{2}-\frac{\pi}{4}+O\left(\frac{1}{a^2\lambda}\right)\right). \tag{7.25}$$

We use Theorem 15 for the function $M_{-i\varkappa,0}(-ia^2\xi)$; letting $\Phi(z)$ denote the function symbolized in that theorem by $\zeta(z)$ and setting $z=\frac{a^2}{4\varkappa}\xi(x,\lambda)$, we have

$$M_{-i\varkappa,0}(-ia^2\xi)\equiv M_{-i\varkappa,0}(-4i\varkappa z)=(-\varkappa)^{1/6}2^{-1/3}e^{-\pi\varkappa+i\pi/4}(\Phi'(z))^{-1/2}\left(Ai(-(-2\varkappa)^{4/3}\Phi(z))+O\left(\frac{1}{\varkappa^{7/6}|\Phi|^{1/4}}\right)\right). \tag{7.26}$$

We recall that $\xi(\lambda,\lambda)=\frac{4\varkappa}{a^2}$ and that the inequality $\xi(x,\lambda)\leq-(1+\delta)\xi(\lambda,\lambda)$ holds for $x \leq \lambda(1-\delta)$; consequently,

$$z \geq \frac{a^2\xi(x,\lambda)}{4\varkappa} \geq 1+\delta.$$

Therefore, $\Phi(z)$ is positive and [according to the definition (3.57)-(3.58)]

$$\Phi^{3/2}=\frac{3}{2}\int_1^z\sqrt{1-\frac{1}{s}}\,ds=-\frac{3}{8}\frac{a^2}{\varkappa}\int_{\xi}^{\frac{4\varkappa}{a^2}}\sqrt{1-\frac{4\varkappa}{a^2s}}\,ds.$$

Comparing these equations with the equation for $\xi(x,\lambda)$, we obtain

$$\frac{2}{3}\Phi^{3/2}=-\frac{a^2}{2\varkappa}\int_x^{\lambda}\sqrt{\frac{\lambda^2-s^2}{1-s^2}}\,ds=-\frac{a^2}{2\varkappa}\left(\int_0^{\lambda}\sqrt{\frac{\lambda^2-s^2}{1-s^2}}\,ds-\eta(x,\lambda)\right). \tag{7.27}$$

Moreover, $\Phi(\Phi')^2=1-\frac{4\varkappa}{a^2\xi}$, and by the equation for $\xi(x,\lambda)$

$$\Phi(\Phi')^2\left(\frac{d\xi}{dx}\right)^2=4\,\frac{\lambda^2-x^2}{1-x^2}=4\left(\frac{d\eta}{dx}\right)^2. \tag{7.28}$$

Together with the asymptotic expansion (3.28) for the Airy function $\mathcal{A}i(z)$, as $z \to -\infty$ these equations yield

$$V(x) = \left(2\pi \frac{d\eta}{dx}\right)^{-1/2} e^{-\pi\varkappa} \cos\left(a^2 \int_0^\lambda \sqrt{\frac{\lambda^2 - s^2}{1-s^2}}\, ds - a^2 \eta\,(x, \lambda) - \frac{\pi}{4} + O\left(\frac{1}{\varkappa}\right)\right). \quad (7.29)$$

If the right-hand sides of (7.25) and (7.29) are asymptotic expansions for one solution, the arguments of the cosines differ only by an integral multiple of π. Consequently, the equation for the eigenvalues in this case has the form

$$\frac{a^2}{\pi} \int_0^{\lambda_n} \sqrt{\frac{\lambda_n^2 - s^2}{1-s^2}}\, ds - \frac{\nu}{2} - \frac{1}{2} + O\left(\frac{1}{\varkappa_n}\right) = n\,, \quad (7.30)$$

where n is an integer. The choice of the phase constant is uniquely determined here by comparing this equation with the corresponding equation derived for $\lambda_n \leq \frac{1}{\sqrt{2}}$. Given the choice made in (7.30), the nth eigenfunction has n and only n zeros in the interval $(0, a)$.

Now, equating the right-hand sides of Eqs. (7.25) and (7.29) at points where the cosines are equal to ± 1, we find the coefficient c_n in the relation $W_n = c_n V_n$:

$$c_n = (-1)^n \frac{2}{a} e^{\pi\varkappa_n} \left(1 + O\left(\frac{1}{\varkappa_n}\right)\right) = (-1)^n \frac{2}{a} \exp\left(-a^2 \int_{\lambda_n}^1 \sqrt{\frac{s^2 - \lambda_n^2}{1-s^2}}\, ds\right)\left(1 + O\left(\frac{1}{\varkappa_n}\right)\right). \quad (7.31)$$

We now consider the case in which $|\varkappa|$ is small relative to a. Equation (7.25) is still valid, and for the Whittaker function it is now required to use the asymptotic equation (3.55) and relation (3.53). This gives

$$M_{-i\varkappa,0}(-ia^2\xi) = \frac{e^{-\pi\varkappa}}{\Gamma(\frac{1}{2}+i\varkappa)} W_{i\varkappa,0}\left(-a^2\xi e^{-i\pi/2}\right) + \frac{e^{-\pi\varkappa + i\pi/2}}{\Gamma(\frac{1}{2}-i\varkappa)} W_{-i\varkappa,0}\left(-a^2\xi e^{i\pi/2}\right) = 2e^{i\pi/4 - \frac{\pi\varkappa}{2}} \operatorname{Re}\left\{\frac{(-a^2\xi)^{i\varkappa} \exp\left(-\frac{ia^2\xi}{2} - \frac{i\pi}{4}\right)}{\Gamma\left(\frac{1}{2}+i\varkappa\right)}\left(1 + O\left(\frac{1+\varkappa^2}{a^2}\right)\right)\right\} \quad (7.32)$$

[here $\arg(-a^2\xi) = 0$]. We observe that (in the notation $z = \frac{a^2\xi}{4\varkappa}$)

$$-\frac{i}{2}a^2\xi + i\varkappa \ln(-a^2\xi) = -2i\varkappa\left(z - \frac{1}{2}\ln 4z\right) + i\varkappa \ln(-\varkappa) = -2i\varkappa \int_1^z \sqrt{1-\frac{1}{s}}\, ds - i\varkappa + i\varkappa \ln(-\varkappa) + O\left(\frac{\varkappa}{z}\right). \quad (7.33)$$

Thus,

$$(-a^2\xi)^{i\varkappa} e^{-\frac{i}{2}a^2\xi} = \left(1 - \frac{4\varkappa}{a^2\xi}\right)^{-1/4} \exp\left(ia^2 \int_\varkappa^\lambda \sqrt{\frac{\lambda^2 - s^2}{1-s^2}}\, ds + i\varkappa \ln(-\varkappa) - i\varkappa\right)\left(1 + O\left(\frac{1+\varkappa^2}{a^2}\right)\right) \quad (7.34)$$

[here again we use the equation for $\xi(x, \lambda)$]. Therefore,

$$V(x) = \frac{e^{-\frac{\pi}{2}\varkappa}}{|\Gamma(\frac{1}{2}+i\varkappa)|} (\eta')^{-\frac{1}{2}} \left\{\cos\left(a^2 \int_0^\lambda \sqrt{\frac{\lambda^2 - s^2}{1-s^2}}\, ds - a^2\eta(x, \lambda) + \varkappa \ln(-\varkappa) - \varkappa - \arg\Gamma\left(\frac{1}{2}+i\varkappa\right) - \frac{\pi}{4}\right) + O\left(\frac{1+\varkappa^2}{a^2}\right)\right\}, \quad (7.35)$$

where in the role of $\arg\Gamma\left(\frac{1}{2}+i\varkappa\right)$ we choose the branch equal to $\varkappa \ln(-\varkappa) - \varkappa + O\left(\frac{1}{\varkappa}\right)$ for $\varkappa \to -\infty$. We note that for sufficiently large values of $|\varkappa|$ Eq. (7.35) goes over to (7.29), provided only that the remainder term is replaced by $O(1/\varkappa)$. Consequently, the eigenvalue equation obtained by comparison of (7.35) and (7.25) can be consolidated with (7.30). Finally, taking account of the result of Lemma 11, we verify the following.

Theorem 24. If the nth eigenvalue λ_n of the boundary-value problem for Eq. (7.2) is not greater than unity, it satisfies the asymptotic equation

$$\frac{a^2}{\pi}\int_0^{\lambda_n}\sqrt{\frac{\lambda_n^2-s^2}{1-s^2}}\,ds+\frac{1}{\pi}\varkappa_n\ln(-\varkappa_n)-\frac{\varkappa_n}{\pi}-\frac{1}{\pi}\arg\Gamma\left(\frac{1}{2}+i\varkappa_n\right)-\frac{\nu}{2}-\frac{1}{2}+q(n,a)=n, \tag{7.36}$$

where $\varkappa_n$ denotes the value of the right-hand side of (7.9) for $\lambda=\lambda_n$ and

$$q(a,n)=O\left(\min\left(\frac{1+n^2}{a^2},\frac{1}{n}\right)\right)+O\left(\min\left(\frac{1+(a^2-n\pi)^2}{a^2},\frac{1}{a^2-n\pi}\right)\right). \tag{7.37}$$

We also note that the following must be written in place of (7.31):

$$C_n=\frac{(-1)^n}{a}\left(\frac{e}{\pi}\right)^{1/2}e^{\frac{\pi}{2}\varkappa_n}\left|\Gamma\left(\frac{1}{2}+i\varkappa_n\right)\right|\left(1+O\left(\min\left(\frac{1}{\varkappa_n},\frac{1+\varkappa_n^2}{a^2}\right)\right)\right). \tag{7.38}$$

7.7. Equation for Large Eigenvalues. We first examine the case in which $\varkappa>0$ and the ratio $\varkappa/a$ is sufficiently large. In this case the asymptotic equation for $M_{i\varkappa,0}(ia^2\xi)$ follows at once from (7.32) by virtue of the Kummer transformation:

$$M_{i\varkappa,0}(ia^2\xi)=e^{-i\pi/2}M_{-i\varkappa,0}(-ia^2\xi),\quad -a^2\xi>0. \tag{7.39}$$

Then, setting $z=-\frac{a^2\xi}{4\varkappa}$, we have [by analogy with (7.33)]

$$-\frac{1}{2}a^2\xi+\varkappa\ln(-a^2\xi)=2\varkappa\int_0^z\sqrt{1+\frac{1}{s}}\,ds+\varkappa\ln\varkappa-\varkappa+O\left(\frac{\varkappa}{z}\right). \tag{7.40}$$

Using the equation for $\xi(x,\lambda)$, we deduce from the above

$$V(x)=\frac{e^{-\frac{\pi}{2}\varkappa}}{|\Gamma(\frac{1}{2}+i\varkappa)|}(\eta')^{-1/2}\cos\left(a^2\int_x^1\sqrt{\frac{\lambda^2-s^2}{1-s^2}}\,ds+\varkappa\ln\varkappa-\varkappa-\arg\Gamma\left(\frac{1}{2}+i\varkappa\right)-\frac{\pi}{4}+O\left(\frac{1+\varkappa^2}{a^2}\right)\right). \tag{7.41}$$

By virtue of Theorem 14, as we can readily verify, this equation is valid for large $\varkappa$ as well if the remainder term is replaced by $O\left(\frac{1}{\varkappa}+\frac{\ln\lambda}{\lambda a^2}\right)$. Comparing the right-hand side of (7.41) with (7.25), we arrive at the following result.

Theorem 25. If the nth eigenvalue λ_n of the boundary-value problem for Eq. (7.2) is not smaller than unity, it satisfies the equation

$$\frac{a^2}{\pi}\int_0^1\sqrt{\frac{\lambda_n^2-s^2}{1-s^2}}\,ds+\frac{1}{\pi}\left(\varkappa_n\ln\varkappa_n-\varkappa_n-\arg\Gamma\left(\frac{1}{2}+i\varkappa_n\right)\right)-\frac{\nu}{2}-\frac{1}{2}+\tilde{q}(a,n)=n, \tag{7.42}$$

in which

$$\tilde{q}(a,n)=O\left(\min\left(\frac{(n\pi-a^2)^2+1}{a^2},\frac{1}{n\pi-a^2}+\frac{\ln\lambda_n}{\lambda_n a^2}\right)\right). \tag{7.43}$$

We note that Eq. (7.38) remains in force for $\varkappa\geq0$; the only difference is that in the event of negative $\varkappa_n$ having a large modulus the modulus of C_n is exponentially small [according to (7.31)], while for $\varkappa_n\to+\infty$

$$C_n=(-1)^n\frac{2}{a}\left(1+O\left(\frac{1}{\varkappa_n}\right)\right). \tag{7.44}$$

7.8. Normalized Eigenfunctions. Here we calculate the normalized integral for the nth eigenfunction. It is equal to

$$a\int_0^{1/2}\frac{W_n^2(x)}{1-x^2}dx+a\,c_n^2\int_{1/2}^{1}\frac{V_n^2(x)}{1-x^2}dx. \tag{7.45}$$

For the calculation we use the asymptotic expansions of the functions W and V. It follows from (7.16) that for $a^2\eta\geqslant 1$

$$W^2=\frac{2}{\pi a^2\eta'(x)}\left(\tfrac{1}{2}+\tfrac{1}{2}\cos\left(2a^2\eta-\nu\pi-\tfrac{\pi}{2}+O\left(\tfrac{1}{a^2\eta}\right)\right)\right). \tag{7.46}$$

Inasmuch as $\eta(x,\lambda)\geqslant\lambda x$ for small x, this equation holds for $x\geqslant\frac{1}{\lambda a^2}$. Moreover, in the interval $0\leqslant x\leqslant\frac{1}{\lambda a^2}$

$$W^2=O\left(\frac{a^{4\nu}\eta^{2\nu+1}}{\lambda}\right)=O\left(a^{4\nu}\lambda^{2\nu}x^{2\nu+1}\right).$$

Therefore,

$$\int_0^{1/2}\frac{W^2(x)}{1-x^2}dx=\frac{1}{\pi a^2}\int_0^{1/2}\left(1+\cos\left(2a^2\eta-\pi\nu-\tfrac{\pi}{2}\right)\right)\frac{dx}{(1-x^2)\eta'}+O\left(\frac{1}{\lambda^2a^4}\int_{1/\lambda a^2}^{1/2}\frac{dx}{x}\right). \tag{7.47}$$

The function $(\eta')^{-1}(1-x^2)^{-1}$ is clearly monotonically increasing; the integral with the cosine can therefore be estimated by means of the theorem of the mean. We then obtain

$$\int_0^{1/2}\frac{W^2}{1-x^2}dx=\frac{1}{\pi a^2}\int_0^{1/2}\frac{dx}{\sqrt{(\lambda^2-x^2)(1-x^2)}}\left(1+O\left(\frac{\ln(\lambda a^2)}{\lambda a^2}\right)\right). \tag{7.48}$$

The calculation of the second term in (7.45) is a bit more complicated. We shall compute this integral only for large values of $|\varkappa_n|$; in the case of small $|\varkappa_n|$, i.e., when it does not exceed a set constant, the calculations are considerably more intricate. We first examine the case of $\lambda_n<1$. In this case, on the grounds of Lemma 13 and Theorem 15,

$$\int_{1/2}^{1}\frac{V_n^2}{1-x^2}dx=2^{-5/3}(-\varkappa_n)^{1/3}e^{-2\pi\varkappa_n}\int_{1/2}^{\lambda_n}(\varphi'\xi')^{-1}A_i^2\left(-(-2\varkappa)^{2/3}\varphi\right)\frac{dx}{1-x^2}\left(1+O\left(\tfrac{1}{\varkappa_n}\right)\right)+O\left(|\varkappa|^{1/3}e^{-2\pi\varkappa_n}\right)\int_{\lambda_n}^{1}(\varphi'\xi')^{-1}A_i^2\left(-(-2\varkappa)^{2/3}\varphi\right)\frac{dx}{1-x^2}. \tag{7.49}$$

For the Airy function we can use the asymptotic expansion if the quantity $|\varkappa^{2/3}\varphi|$ is large enough, i.e., [by (7.27)], if $a^2|\lambda-x|^{3/2}\geqslant\sqrt{1-\lambda}$. In the interval in which this inequality does not hold the Airy function is bounded, and $(\varphi'\xi')^{-1}$ is estimated, due to (7.28) and (7.27), by the quantity $O(a^{4/3}\varkappa^{-1})$. Thus, the right-hand side of (7.49) is equal to

$$\frac{1}{2\pi}e^{-2\pi\varkappa_n}\int_{\frac{1}{2}}^{\lambda-\varepsilon}\left(1+O\left(\frac{1}{\varkappa\varphi^{3/2}}\right)\right)\cos^2\left(a^2\eta-\frac{\nu\pi}{2}-\frac{\pi}{4}\right)\frac{dx}{(1-x^2)\eta'}+O\left(|\varkappa|^{-1/3}\right)+O\left(e^{-2\pi\varkappa_n}\right)\int_{\lambda+\varepsilon}^{1}\exp\left(-a^2\int_\lambda^x\sqrt{\frac{s^2-\lambda^2}{1-s^2}}ds\right)\frac{dx}{(1-x^2)\eta'}, \tag{7.50}$$

where $\varepsilon=a^{-4/3}(1-\lambda)^{1/3}$. Furthermore,

$$|\varkappa\varphi^{3/2}|\geqslant\frac{a^2}{8}\left|\int_\lambda^x\left|\frac{\lambda-s}{1-s}\right|^{1/2}ds\right|\geqslant\frac{a^2}{16}\frac{|\lambda-x|^{3/2}}{\sqrt{1-\lambda}}.$$

Substituting this inequality into (7.50), we obtain

$$\int_{\frac{1}{2}}^{\lambda-\varepsilon}\frac{dx}{\varkappa\varphi^{3/2}(1-x^2)\eta'}=O\left(\int_{\frac{1}{2}}^{\lambda-\varepsilon}\frac{dx}{a^2(\lambda-x)^2}\right)=O\left(\frac{1}{a^2\varepsilon}\right)=O\left(|\varkappa|^{-1/3}\right). \tag{7.51}$$

Estimating the cosine integral as before, we find

$$\int_{\frac{1}{2}}^{1} \frac{V^2(x)}{1-x^2}\,dx = \frac{e^{-2\pi\varkappa_n}}{4\pi}\int_{1/2}^{\lambda_n} \frac{ds}{\sqrt{(\lambda_n^2-s^2)(1-s^2)}}\left(1+O\left(\frac{1}{|\varkappa|^{1/3}}\right)\right). \tag{7.52}$$

Consequently, if the eigenfunction is normalized so as to make it coincide with the solution $W(x)$ for small x, its normalized integral is equal to

$$D_n^2 \equiv a\int_0^1 \frac{w_n^2(x)}{1-x^2}\,dx = \frac{2}{\pi a}\int_0^{\lambda_n} \frac{ds}{\sqrt{(\lambda_n^2-s^2)(1-s^2)}}\left(1+O\left(\frac{1}{|\varkappa_n|^{1/3}}\right)\right). \tag{7.53}$$

We employ precisely the same procedure to ascertain the following for sufficiently large positive $\varkappa$:

$$D_n^2 = \frac{2}{\pi a}\int_0^1 \frac{ds}{\sqrt{(\lambda_n^2-s^2)(1-s^2)}}\left(1+O\left(\frac{1}{\varkappa_n}\right)\right). \tag{7.54}$$

Now by simply combining the foregoing estimates we obtain the following asymptotic equations for the eigenfunctions of the boundary-value problem (7.1).

Theorem 26. Let $\psi_n(x,a)$ be the nth eigenfunction of problem (7.1), and let its corresponding eigenvalue λ_n be no smaller than $\frac{1}{\sqrt{2}}$. Then for sufficiently large a and $\lambda_n \leq 1$ (in the notation of Lemmas 12 and 13)

$$D_n\psi_n(ax,a) = \begin{cases} \left(\dfrac{\eta}{(1-x^2)\eta'}\right)^{+\frac{1}{2}}\left(J_\nu(a^2\eta) + O\left(\dfrac{a^{2\nu}\eta^{\nu+2}}{\lambda_n(1+a^2\eta)(1+(a^2\eta)^{\nu+1/2})}\right)\right), & 0 \leq x \leq \frac{1}{\sqrt{2}}-\delta; \quad (7.55)\\ (-1)^n a^{-1}\left(\dfrac{2}{\pi}\right)^{1/2} e^{\frac{\pi}{2}\varkappa_n}\left|\Gamma\left(\tfrac{1}{2}+i\varkappa_n\right)\right|\left((1-x^2)\xi'(x,\lambda_n)\right)^{-\frac{1}{2}}\Big(\operatorname{Re} M_{-i\varkappa_n,0}(-ia^2\xi) + & \\ +O\left(\dfrac{\ln a}{a^2}\right)\beta(\xi,\varkappa_n)e^{2\varkappa_n\vartheta\left(\frac{a^2\xi}{4\varkappa_n}\right)}\Big), & \frac{1}{4} \leq x < +\infty; \quad (7.56)\end{cases}$$

and for $\lambda_n \geq 1$ the latter equation must be replaced by

$$D_n\psi_n(ax,a) = (-1)^n a^{-1}\left(\frac{2}{\pi}\right)^{1/2} e^{\frac{\pi}{2}\varkappa_n}\left|\Gamma\left(\tfrac{1}{2}+i\varkappa_n\right)\right|\left((1-x^2)\xi'(x,\lambda_n)\right)^{-1/2}\Big(\operatorname{Re} M_{i\varkappa_n,0}(ia^2\xi) + \\ +O\left(\frac{\ln a\,\ln(\lambda_n+1)}{\lambda_n a^2}\right)\beta(\xi,\varkappa_n)e^{2\varkappa_n\vartheta\left(\frac{a^2\xi}{4\varkappa_n}\right)}\Big), \quad \frac{1}{4} \leq x < +\infty. \tag{7.57}$$

7.9. Asymptotic Equations for the Eigenfunctions When $n \to \infty$. The asymptotic equations of Theorem 26 are simplified somewhat when the ratio n/a^2 is large. In this case we can use the asymptotic expansions of Theorems 14 and 15 in Sec. 3.6 for the Whittaker functions. We infer the following as a result.

Lemma 14. Let the quantity n/a^2 be sufficiently large; then the following asymptotic equations are valid for the nth eigenfunction of the boundary-value problem (7.1):

$$\frac{\psi_n(ax,a)}{\psi_n(a,a)} = \begin{cases} (-1)^n \eta^{1/2}\left((1-x^2)(\lambda_n^2-x^2)\right)^{-1/4}\left(J_\nu(a^2\eta(x,\lambda_n)) + O\left(\frac{1}{\lambda_n a^2(1+a\sqrt{\eta})}\right)\right), \quad 0 \le x \le \frac{1}{2}; & (7.58) \\ \eta_1^{1/2}\left((1-x^2)(\lambda_n^2-x^2)\right)^{-1/4}\left(J_0(a^2\eta_1(x,\lambda_n)) + O\left(\frac{1}{\lambda_n a^2(1+a\sqrt{\eta_1})}\right)\right), \quad \frac{1}{2} \le x \le 1; & (7.59) \\ \eta_2^{1/2}\left((x^2-1)(\lambda_n^2-x^2)\right)^{-\frac{1}{4}}\left(I_0(a^2\eta_2(x,\lambda_n)) + O\left(\frac{\ln^2\lambda_n e^{a^2\eta_2(x,\lambda_n)}}{\lambda_n a^2(1+a\sqrt{\eta_2})}\right)\right), \quad 1 \le x \le \frac{\lambda_n}{2}; & (7.60) \\ 2^{1/2} a^{-\frac{2}{3}} \exp\left(a^2 \int_1^{\lambda_n} \sqrt{\frac{\lambda_n^2-s^2}{s^2-1}}\, ds\right)\left((1-x^2)\xi'\right)^{-1/2}\left(Ai(-a^{4/3}\xi(x,\lambda_n)) + O\left(\frac{\ln^2\lambda_n \left|\exp\left(-\frac{2}{3}a^2(-\xi)^{3/2}\right)\right|}{\lambda_n a^2(1+a^{1/3}|\xi|^{1/4})}\right)\right), \quad \lambda_{n/2} \le x < +\infty, & (7.61) \end{cases}$$

where

$$\eta(x,\lambda) = \int_0^x \sqrt{\frac{\lambda^2-s^2}{1-s^2}}\, ds, \quad \eta_1(x,\lambda_n) = \eta(1,\lambda) - \eta(x,\lambda), \quad \eta_2(x,\lambda) = \int_1^x \sqrt{\frac{\lambda^2-s^2}{s^2-1}}\, ds, \tag{7.62}$$

and

$$\xi(x,\lambda) = \begin{cases} -\left(\frac{3}{2}\int_x^\lambda \sqrt{\frac{\lambda^2-s^2}{s^2-1}}\, ds\right)^{2/3}, & 1 \le x \le \lambda; \\ \left(\frac{3}{2}\int_\lambda^x \sqrt{\frac{s^2-\lambda^2}{s^2-1}}\, ds\right)^{2/3}, & \lambda \le x < \infty. \end{cases} \tag{7.63}$$

We note first of all that for large λ_n

$$e^{\frac{\pi}{2}\varkappa_n}\left|\Gamma\left(\tfrac{1}{2}+i\varkappa_n\right)\right| = \sqrt{2\pi} + O\left(\frac{1}{\varkappa_n}\right) = \sqrt{2\pi} + O\left(\frac{1}{\lambda_n a^2}\right).$$

Moreover, when $x \to 1$,

$$\frac{\xi(x)}{x^2-1}\left(\frac{d\xi}{dx}\right)^{-1} = \frac{1}{2} + o(1).$$

Since, in addition, $M_{i\varkappa,0}(ia^2\xi) = e^{i\pi/4} a \xi^{1/2}(1+o(\xi))$ when $\xi \to +0$ it follows from (7.57) that

$$\psi_n(a,a) = \frac{(-1)^n}{\mathcal{D}_n}\left(1 + O\left(\frac{\ln a \ln \lambda_n}{\lambda_n a^2}\right)\right). \tag{7.64}$$

We immediately deduce (7.58) from (7.64) and (7.55). Then, letting $\phi(z)$ denote the function symbolized in Theorem 14 by $\eta(z)$, we have the following for $x>1$ on the grounds of Eq. (7.7):

$$2\varkappa_n \phi\left(\frac{a^2\xi(x,\lambda_n)}{4\varkappa_n}\right) = 2\varkappa_n \int_0^{\frac{a^2\xi}{4\varkappa_n}} \left(\frac{1}{s}-1\right)^{1/2} ds = a^2 \int_1^x \sqrt{\frac{\lambda_n^2-s^2}{s^2-1}}\, ds. \tag{7.65}$$

For $x>1$, therefore, the function $2\varkappa_n\phi\left(\frac{a^2\xi(x,\lambda_n)}{4\varkappa_n}\right)$ coincides with $a^2\eta(x,\lambda_n)$; for $x<1$ this function is equal to $-ia^2\eta_1(x,\lambda_n)$. Since $I_0(\pm iz) = J_0(z)$, (3.78) and (7.56) lead to Eqs. (7.59) and (7.60) when $\xi \le 4(1-\delta)\varkappa_n$; this condition is clearly fulfilled for $x \le \frac{1}{2}\lambda_n$.

Finally, (7.61) follows in exactly the same way from Theorem 15 and the equation for the function $\xi(x,\lambda_n)$.

§ 8. Asymptotic Equations for the Eigenvalues of the Integral Equation

We conclude our investigation of the integral equation (5.1) by deriving asymptotic equations for the large-order eigenvalues of that equation.

Theorem 27. Let μ_n be the nth eigenvalue of the integral equation (5.1), and let $n \geq A_0 a^2$. Then

$$\mu_n = \begin{cases} \frac{(-1)^n}{\sqrt{2\pi}} e^{-\frac{\pi}{2}\varkappa_n} \left|\Gamma\left(\frac{1}{2}+i\varkappa_n\right)\right| \left(1+O\left(\frac{\ln a}{a^2}\right)\right), \ \varkappa_n \leq 0; & (8.1) \\ \frac{(-1)^n}{\sqrt{2\pi}} e^{-\frac{\pi}{2}\varkappa_n} \left|\Gamma\left(\frac{1}{2}+i\varkappa_n\right)\right| \left(1+O\left(\frac{\ln a \ln(1+\lambda_n)}{\lambda_n a^2}\right)\right), \ \varkappa_n \geq 0, & (8.2) \end{cases}$$

where $\varkappa_n$ is determined by Eqs. (7.9)-(7.10) for $\lambda=\lambda_n$.

We first examine the case $\lambda_n \geq 1$. From the integral equation

$$\mu_n \Psi_n(x,a) = \int_0^a \sqrt{x\xi}\, J_\nu(x\xi)\, \Psi_n(\xi)\, d\xi \tag{8.3}$$

we arrive immediately at the following when $x \to +\infty$:

$$\mu_n \Psi_n(x,a) = \frac{\sqrt{a}\,\Psi_n(a,a)}{\sqrt{x}} J_{\nu+1}(ax) + O\left(\frac{1}{x^2}\right) = \left(\frac{2}{\pi}\right)^{1/2} \Psi_n(a,a)\left(\frac{\cos\left(ax - \frac{\nu\pi}{2} - \frac{3\pi}{4}\right)}{x} + O\left(\frac{1}{x^2}\right)\right) \tag{8.4}$$

[here we have incorporated the recursive relation $z^{\nu+1} J_\nu(z) = \frac{d}{dz} z^{\nu+1} J_{\nu+1}(z)$ and then the asymptotic expansion of the Bessel function]. On the other hand, it follows from (7.57) that when $x \to +\infty$

$$\Psi_n(ax,a) = \frac{2^{1/2}\Psi_n(a,a)}{ax}\left(1+O\left(\frac{1}{x}\right)\right)\left(\operatorname{Re} M_{i\varkappa,0}\left(ia^2\xi(x,\lambda_n)\right) + O\left(\frac{e^{\pi\varkappa_n}\ln a \ln(1+\lambda_n)}{\lambda_n a^2}\right)\right), \tag{8.5}$$

in which the constant is obtained from (7.57) with $x \to +a$. Arguing exactly as in the derivation of (7.35), we find for $x \to +\infty$

$$\operatorname{Re} M_{i\varkappa,0}\left(ia^2\xi(x,\lambda_n)\right) = \frac{2^{1/2} e^{\frac{\pi}{2}\varkappa_n}}{\left|\Gamma\left(\frac{1}{2}+i\varkappa_n\right)\right|}\cos\left(a^2\int_{\lambda_n}^{x}\sqrt{\frac{s^2-\lambda_n^2}{s^2-1}}\,ds + \arg\Gamma\left(\frac{1}{2}+i\varkappa_n\right) - \varkappa_n \ln\varkappa_n + \varkappa_n - \frac{\pi}{4} + O\left(\frac{\ln a \ln(\lambda_n+1)}{\lambda_n a^2}\right)\right). \tag{8.6}$$

Moreover, when $x \to \infty$,

$$\int_\lambda^x \sqrt{\frac{s^2-\lambda^2}{s^2-1}}\,ds = x - \lambda + \int_\lambda^\infty \left(\sqrt{\frac{s^2-\lambda^2}{s^2-1}} - 1\right) ds + O\left(\frac{1}{x}\right). \tag{8.7}$$

The derivative of the integral on the right-hand side with respect to λ is equal to

$$1 - \lambda\int_\lambda^\infty \frac{ds}{\sqrt{(s^2-\lambda^2)(s^2-1)}} = 1 - \lambda\int_0^1 \frac{ds}{\sqrt{(\lambda^2-s^2)(1-s^2)}} = 1 - \frac{\partial}{\partial\lambda}\int_0^1 \sqrt{\frac{\lambda^2-s^2}{1-s^2}}\,ds, \tag{8.8}$$

and for $\lambda=1$ this integral is equal to zero. Consequently,

$$\int_{\lambda_n}^\infty \left(\sqrt{\frac{s^2-\lambda_n^2}{s^2-1}} - 1\right) ds = \lambda_n - \int_0^1 \sqrt{\frac{\lambda_n^2-s^2}{1-s^2}}\,ds. \tag{8.9}$$

Thus, the argument of the cosine in (8.6) is equal to $a^2 x - \frac{\nu\pi}{2} - \frac{3\pi}{4} - n\pi$ [on account of Eq. (7.42)], and a comparison of (8.6) and (8.4) leads to Eq. (8.2).

The proof of Eq. (8.1) is identical. We note that for large $|\varkappa_n|$ when $\varkappa_n < 0$ the following is implied by (8.1) (with the aid of the Stirling expansion):

$$\mu_n = (-1)^n + O\left(\frac{1}{\varkappa_n}\right). \tag{8.10}$$

But if $\varkappa_n \to +\infty$ then

$$\mu_n = (-1)^n e^{-\pi\varkappa_n}\left(1 + O\left(\frac{\ln a \ln \lambda_n}{\lambda_n a^2}\right)\right) = (-1)^n \exp\left(-a^2 \int_1^{\lambda_n} \sqrt{\frac{\lambda_n^2 - s^2}{s^2 - 1}}\, ds\right)\left(1 + O\left(\frac{\ln a \ln \lambda_n}{\lambda_n a^2}\right)\right). \tag{8.11}$$

Indeed, the remainder term in (8.10) can be substantially improved [to $O(\exp(-\delta|\varkappa_n|))$], but we shall not prove this fact.

Appendix: An Integral Equation with Half-Order Bessel Functions as Its Kernel

We now give some results of an asymptotic investigation of two integral equations:

$$\lambda\psi(x) = \int_0^a \sqrt{x\xi}\, J_{\frac{1}{2}}(x\xi)\psi(\xi)d\xi; \tag{A.1}$$

$$\lambda\psi(x) = \int_0^a \sqrt{x\xi}\, J_{-\frac{1}{2}}(x\xi)\psi(\xi)d\xi. \tag{A.2}$$

Inasmuch as $J_{1/2}(z) = \left(\frac{2}{\pi z}\right)^{1/2} \sin z$ and $J_{-\frac{1}{2}}(z) = \left(\frac{2}{\pi z}\right)^{1/2} \cos z$, the eigenfunctions of Eqs. (A.1) and (A.2) are entire functions of the complex variable x. Each of the eigenfunctions of Eq. (A.1) is an odd, while each of those for Eq. (A.2) is an even function of x. These functions are therefore even and odd eigenfunctions, respectively, of the integral equation

$$\lambda\psi(x) = \frac{1}{\sqrt{2\pi}} \int_{-a}^{a} e^{ix\xi}\psi(\xi)d\xi. \tag{A.3}$$

The eigenfunctions of Eq. (A.1) in this case correspond to the real eigenvalues of Eq. (A.3), and the eigenfunctions of Eq. (A.2) correspond to the imaginary eigenvalues of Eq. (A.3). We denote the eigenfunctions of (A.3) by $\psi_n(x,a)$, enumerating them in order of increasing number of zeros in the interval $(-a,a)$. We denote the eigenvalues of Eq. (A.3) by $i^n\mu_n$, separating out the factor i^n from the nth eigenvalue. The numbers μ_n satisfy the conditions (see [5])

$$1 > \mu_0 > \mu_1 > \ldots > \mu_n > \ldots > 0. \tag{A.4}$$

The functions $\psi_n(x,a)$ are real when x is real (see [5]) and are eigenfunctions of the boundary-value problem (see [5])

$$\frac{d}{dx}(x^2 - a^2)\frac{d}{dx}\psi + a^2x^2\psi = a^4\lambda^2\psi, \qquad |\psi(\pm a)| < \infty, \tag{A.5}$$

where the spectral parameter λ may be assumed positive. Consequently, for the analysis of the asymptotic behavior of these functions in the case of large a we can reiterate verbatim the same arguments we used earlier. Moreover, in the interval $|x| \geq 1$ the principal terms of the asymptotic equations for the solutions of Eq. (A.5) coincide with the principal terms of the corresponding asymptotic expressions of the previously analyzed equation with $\nu > \frac{1}{2}$ (except for the provision that λ is not too small). The only difference arises in the case of small λ. In this case it is required to approximate the solutions of Eq. (A.5) by Weber functions (rather than Whittaker functions). The corresponding theorem on approximation by Weber functions has been proved in [31] (see also [17]; we shall use the notation of [31]). We therefore omit the details of the analysis and give only the final result.

First, we write the equations for the eigenvalues of the boundary-value problem (A.5). We denote these eigenvalues by λ_n (they are all positive and enumerated in increasing order). We put

$$\chi(v) = v\ln v - v + \arg\Gamma\left(\frac{1}{2} - iv\right) \tag{A.6}$$

[we pick the branch of $\arg \Gamma\left(\frac{1}{2}-iv\right)$ that is equal to $-v \ln v + v + O\left(\frac{1}{v}\right)$ when $v \to +\infty$] and let

$$\gamma = \frac{2}{\pi} \int_{\lambda}^{1} \left(\frac{s^2-\lambda^2}{1-s^2}\right)^{1/2} ds, \quad \lambda \leq 1; \tag{A.7}$$

$$\gamma = -\frac{2}{\pi} \int_{1}^{\lambda} \left(\frac{\lambda^2-s^2}{s^2-1}\right)^{1/2} ds, \quad \lambda \geq 1. \tag{A.8}$$

In this notation the equations for the eigenvalues λ_n have the following form.

Theorem 1. If the nth eigenvalue λ_n of problem (A.5) is not greater than unity, it satisfies the equation

$$\frac{a^2}{\pi} \int_{0}^{\lambda_n} \sqrt{\frac{\lambda_n^2-s^2}{1-s^2}}\, ds - \chi\left(\frac{1}{2}a^2\gamma(\lambda_n)\right) + O\left(\min\left(\frac{1}{\gamma(\lambda_n)a^2}, \frac{1+\gamma^2(\lambda_n)a^4}{a^2}\right)\right) = \frac{n}{2} + \frac{1}{4}. \tag{A.9}$$

But if $\lambda_n \geq 1$, it satisfies the equation

$$\frac{a^2}{\pi} \int_{0}^{1} \sqrt{\frac{\lambda_n^2-s^2}{1-s^2}}\, ds + \chi\left(-\frac{1}{2}a^2\gamma(\lambda_n)\right) + O\left(\min\left(\frac{1}{|\gamma(\lambda_n)|a^2}, \frac{1+\gamma^2(\lambda_n)a^4}{a^2}\right)\right) = \frac{n}{2} + \frac{1}{4}. \tag{A.10}$$

The following holds true for the eigenvalues of the integral equation (A.3).

Theorem 2. For sufficiently large a

$$\mu_n = \frac{1}{\sqrt{2\pi}} \exp\left(\frac{\pi a^2}{4}\gamma(\lambda_n)\right) \left|\Gamma\left(\frac{1}{2} + \frac{ia^2}{4}\gamma(\lambda_n)\right)\right| \left(1 + O\left(\min\left(\frac{1}{|\gamma(\lambda_n)|a^2}, \frac{1}{a^2} + a^2\gamma^2(\lambda_n)\right)\right)\right). \tag{A.11}$$

As in Theorem 23, the remainder term can be greatly improved for small n and an exponential estimate obtained for the difference $1-\mu_n$. This has been done, nonuniformly on n, by Fuchs in [5]. Widom [32] has obtained an estimate analogous to (A.11) for $n \to \infty$ (though without remainders).

We write the asymptotic equations for the eigenfunctions of Eq. (A.3) only for two cases, namely: when λ_n satisfies either the inequality $0 < \lambda_n \leq \delta$ or the inequality $\delta \leq \lambda_n \leq 1-\delta$. We define the function $\xi(x, \lambda)$ by the equation

$$(\xi^2 - c^2)\left(\frac{d\xi}{dx}\right)^2 = \frac{x^2-\lambda^2}{1-x^2}, \tag{A.12}$$

in which

$$c^2 = \frac{4}{\pi} \int_{0}^{\lambda} \sqrt{\frac{\lambda^2-s^2}{1-s^2}}\, ds,$$

and by the auxiliary conditions $\xi(\pm\lambda, \lambda) = \pm c$. Also, let

$$2\sqrt{\eta(x,\lambda)} = \int_{x}^{1} \sqrt{\frac{s^2-\lambda^2}{1-s^2}}\, ds, \tag{A.13}$$

where $\arg \eta = 0$ for $x < 1$ and $\arg \eta = \pi$ for $x > 1$. Finally, we set

$$\frac{2}{3}(v(x,\lambda))^{3/2} = \int_{\lambda}^{x} \sqrt{\frac{s^2-\lambda^2}{1-s^2}}\, ds, \tag{A.14}$$

in which $v>0$ for $\lambda<x<1$ and $v<0$ for $0<x<\lambda$. We introduce the additional notation

$$\mathcal{D}_n = \left(\frac{2n!}{\sqrt{\pi}} \int_0^{\lambda_n} \frac{ds}{\sqrt{(\lambda_n^2 - s^2)(1-s^2)}}\right)^{-1/2}; \tag{A.15}$$

$$\tilde{\mathcal{D}}_n = \left(\frac{1}{\pi} \int_0^{\lambda_n} \frac{ds}{\sqrt{(\lambda_n^2 - s^2)(1-s^2)}}\right)^{-1/2}. \tag{A.16}$$

In this notation the asymptotic equations for the eigenfunctions of Eq. (A.3) acquire the following form.

Theorem 3. Let n be such that $\lambda_n \leq 2^{-6}$. Then

$$\psi_n(ax,a) = \begin{cases} \mathcal{D}_n\left((1-x^2)\xi'(x,\lambda_n)\right)^{-1/2}\left(U(-n-\tfrac{1}{2}, a\sqrt{2}\,\xi(x,\lambda_n)) + R_1\right), & 0 \leq x \leq 1-2^{-6} \quad (A.17) \\ 2^{-1/4}(a\pi)^{1/2}\left(n+\tfrac{1}{2}\right)^{\frac{n}{2}+\frac{1}{4}} e^{-\frac{n}{2}-\frac{1}{4}-\frac{\pi a^2}{2}\gamma(\lambda_n)} \mathcal{D}_n \left(\frac{\eta(x,\lambda_n)}{(x^2-1)(x^2-\lambda_n^2)}\right)^{1/4}\left(J_0(2a^2\sqrt{\eta(x,\lambda_n)}) + R_2\right), & 1-2^{-6} \leq x < +\infty, \quad (A.18) \end{cases}$$

where $U(p,z)$ is the Weber function and, with the constant κ independent of a and n,

$$|R_1| \leq \kappa \left(\frac{n+1}{e}\right)^{\frac{n}{2}+\frac{1}{4}} \left|\exp\left(-(2n+1)\int_1^{\frac{ax}{\sqrt{2n+1}}} \sqrt{1-t^2}\,dt\right)\right| \left(1+n^{1/12}+a\left|x^2-\frac{2n+1}{a^2}\right|^{1/4}\right)^{-1} \frac{\ln a}{a^3}, \tag{A.19}$$

$$|R_2| \leq \kappa \frac{1}{1+a|\eta|^{1/4}} \min\left(\frac{1}{n}, \frac{1+n^2}{a^2}\right) \left|e^{2ia^2\sqrt{\eta(x,\lambda_n)}}\right|. \tag{A.20}$$

Theorem 4. Let n be such that $2^{-6} \leq \lambda_n \leq 1-2^{-6}$. Then

$$\psi_n(ax,a) = \begin{cases} \tilde{\mathcal{D}}_n a^{-1/6}\left(\frac{v(x,\lambda_n)}{(1-x^2)(x^2-\lambda_n^2)}\right)^{1/4}\left(Ai(a^{4/3}v(x,\lambda_n)) + R_3\right), & 0 \leq x \leq 1-2^{-7}; \quad (A.21) \\ \tilde{\mathcal{D}}_n a^{1/2} e^{-\frac{\pi a^2}{2}\gamma(\lambda_n)}\left(\frac{\eta(x,\lambda_n)}{(x^2-1)(x^2-\lambda_n^2)}\right)^{1/4}\left(J_0(2a^2\sqrt{\eta(x,\lambda_n)}) + R_4\right), & 1-2^{-7} \leq x < +\infty, \quad (A.22) \end{cases}$$

where

$$R_3 = O\left(\frac{\left|\exp\left(-\frac{2}{3}a^2 v^{3/2}(x,\lambda_n)\right)\right|}{a^2(1+a^{1/3}|v|^{1/4})}\right), \tag{A.23}$$

$$R_4 = O\left(\frac{\left|\exp(2ia^2\sqrt{\eta(x,\lambda_n)})\right|}{a^2(1+a|\eta|^{1/4})}\right). \tag{A.24}$$

The asymptotic equations of Theorems 3 and 4 are given for the eigenfunctions normalized by the condition $\int_{-a}^{a} \psi_n^2(x,a)\,dx = 1$.

LITERATURE CITED

1. Courant, R., and Hilbert, D., Methods of Mathematical Physics [Russian translation], Vol. 1, Moscow-Leningrad (1951) [English edition: Wiley, New York (1953)].
2. Gel'fond, A. O., On the growth of the eigenvalues of integral equations, in: Lovitt, W. V., Linear Integral Equations [Russian translation], Moscow (1957) [English edition: Dover, New York (1950)].
3. Titchmarsh, E. C., Introduction to the Theory of Fourier Integrals [Russian translation], Moscow-Leningrad (1948) [English edition: Oxford University Press, Cambridge (1937)].

4. Bateman, H., and Erdélyi, A., Higher Transcendental Functions [Russian translation], Vol. 3, Moscow (1967) [English edition: McGraw-Hill, New York (1953)].
5. Fuchs, W., On the eigenvalues of an integral equation, J. Math. Anal. Appl., Vol. 9, No. 3 (1964).
6. Whittaker, E. T., and Watson, G. N., Modern Analysis [Russian translation], Vol. 2, Moscow (1963) [English edition: Cambridge University Press (1927)].
7. Coddington, E. A., And Levinson, N., Theory of Ordinary Differential Equations [Russian translation], IL, Moscow (1958) [English edition: McGraw-Hill, New York (1955)].
8. Titchmarsh, E. C., Eigenfunction Expansions Associated with Second-Order Differential Equations [Russian translation], Vol. 1, IL, Moscow (1960) [English edition: Clarendon, Oxford (1946, 1958)].
9. Watson, G. N., Theory of Bessel Functions [Russian translation], Vol. 1, IL, Moscow (1949) [English edition: Macmillan, New York (1945)].
10. Liouville, J., Second mémoire sur le développement des fonctions ou parties de fonctions en séries dont diverse termes sont assugetis à satisfaire à une même équation différentielle du second ordre contenant un parametre variable, J. Math. Pure et Appl., 2:16-35 (1837).
11. Green, G., On the motion of waves in a variable canal of small depth and width, Trans. Cambridge Phil. Soc., 6:457-462 (1837).
12. Blumenthal, O., Über asymptotische Integration linearer Differentialgleichungen, Arch. Math. Phys. (Leipzig), 19:136-174 (1912).
13. Olver, F., Error bounds for the Liouville–Green approximation, Proc. Cambridge Phil. Soc., Vol. 54, No. 4 (1961).
14. Erdelyi, A., Asymptotic Solutions of Ordinary Linear Differential Equations, Math. Dept., Calif. Inst. Technology, Pasadena, Calif. (1961).
15. Langer, R., On the asymptotic solutions of ordinary differential equations, Trans. Amer. Math. Soc., 33:23-64 (1931).
16. Olver, F., Error bounds for asymptotic expansions in turning-point problems, J. Soc. Indust. Appl. Math., 12:200-214 (1964).
17. Kuznetsov, N. V., Candidate's Dissertation, Matem. Inst., Akad. Nauk SSSR (MIAN) (1965).
18. Miller, J., The Airy Integral, British Association for the Advancement of Science (BAAS), Mathematical Tables, Part-Volume B, Cambridge University Press (1946).
19. Olver, F., Uniform asymptotic expansions for Bessel's functions, Phil. Trans. Roy. Soc. (London), Vol. A247, No. 247 (1954).
20. Erdélyi, A., and Swenson, C., Asymptotic forms of Whittaker's confluent hypergeometric functions, Mem. Amer. Math. Soc. (Providence, R. I.), No. 25 (1967).
21. Vainshtein, L. A., in: High-Power Electronics, Vol. 4, Nauka (1965), pp. 106-134.
22. Slepian, D., Prolate spheroidal wave functions (IV), Bell System Tech. J., 43(6):3009-3057 (1964).
23. Fox, A. G., and Li, T., Resonant modes in a maser interferometer, Bell System Tech. J., 40(2): 453-488 (1961).
24. Boyd, D. G., and Kogelnik, H., Generalized confocal resonator theory, Bell System Tech. J., 41(4):1347-1369 (1962).
25. Slepian, D., Some asymptotic expansions for prolate spheroidal wave functions, J. Math. Phys., Vol. 44, No. 2 (1965).
26. Slavyanov, S. Yu., Asymptotic representations for prolate spheroidal harmonics, Zh. Vychis. Matem. i Matem. Fiz., 7(5):1001-1010 (1967).
27. Skovgaard, H., Uniform Asymptotic Expansions of Confluent Hypergeometric Functions and Whittaker Functions, Copenhagen (1966).
28. Buchholz, H., Die konfluente hypergeometrische Funktion, Berlin (1953).
29. Szego, G., Orthogonal Polynomials (AMS Colloquium Publs. Vol. 23), Amer. Math. Soc., Providence, R. I. (1959).
30. Bateman, H., and Erdélyi, A., Higher Transcendental Functions, McGraw-Hill, New York (1953).
31. Kuznetsov, N. V., Asymptotic distribution of the eigenfrequencies of a plane membrane, Differential'nye Uravneniya, Vol. 1, No. 1 (1966).
32. Widon, H., On the eigenvalues of certain integral equations, J. Math. Pure and Appl. (1964).

SURFACE-WAVE EXCITATION IN CONNECTION WITH DIFFRACTION AT AN IMPEDANCE CONTOUR

I. A. Molotkov

Let S be a sufficiently smooth plane curve with radius of curvature $\rho(s)\neq 0$ (s is the arclength on S measured from some reference point), let an oscillation source be located at the point $M_0(x_0, y_0)$, and let the observation point be located at $M(x, y)$. The points M_0 and M are assumed to be on the same side of the curve S. We wish to investigate the point-source field (Green function) $\Gamma(M_0, M, \kappa)$:

$$(\Delta+K^2)\Gamma(M_0, M, K^2)=-\delta(M-M_0)=-\delta(x-x_0)\delta(y-y_0), \quad (\kappa>0),$$

$$\left(\frac{\partial}{\partial n}+i\kappa g(s)\right)\Gamma\Big|_S=0, \tag{1}$$

$$\lim_{x^2+y^2\to\infty}\Gamma(M_0, M, \kappa)=0 \quad (\mathrm{Im}\,\kappa>0).$$

Here n denotes the normal to S ($n>0$ on the side on which the points M_0 and M are situated), κ is the wave number (which we assume for simplicity is constant), δ is the Dirac delta function, and $g^{-1}(s)$ is a sufficiently smooth complex-valued function, i.e., the normal impedance of the contour S. We assume that

$$\mathrm{Im}\,g(s)<0, \qquad \mathrm{Re}\,g(s)\geq 0, \tag{2}$$

and

$$g(s)\underset{\kappa\to\infty}{=}O(1). \tag{3}$$

Let us suppose that when $\kappa\to\infty$ the Green function can be represented by the sum

$$\Gamma(M_0, M, \kappa)=\Gamma_0(M_0, M, \kappa)+\tilde{\Gamma}(M_0, M, \kappa), \tag{4}$$

in which the term $\Gamma_0(M_0, M, \kappa)$ has the sense of a surface wave, i.e, decays exponentially along n with a coefficient proportional to κ. Our primary concern in the present article is this term $\Gamma_0(M_0, M, \kappa)$, which describes the excitation of a surface wave by a point source.

An enormous number of papers have been published on the propagation of surface waves in connection with the impedance characterization of an interface (see, e.g., the survey [1]). Of the more recent works Grimshaw's paper [2] is important; in it the author investigates the three-dimensional problem of surface wave propagation along a smooth impedance surface. However, the majority of the papers fail to touch on the problem of surface-wave excitation. The excitation of surface waves at curved boundaries has been analyzed so far in [3] and [4]. In [3] the case of constant $\rho(s)$ and $g(s)$

is investigated on the assumption that $g = O(\kappa^{-1/3})$, so that the surface character of this wave is rather feeble. Also lacking in [4] are equations for variable radii of curvature and impedances. Similarly, the asymptotic representation of the surface-wave field is not calculated with acceptable accuracy in [4], so that some of the final equations in that paper contain fallacies.

The objective of the present study is to deduce an asymptotic representation as $\kappa \to \infty$ for the function $\Gamma_0(\mathcal{M}_0, \mathcal{M}, \kappa)$, given arbitrary smooth functions $\rho(s)$ and $g(s)$ satisfying conditions (2) and (3).

§ 1. Surface Wave in the Standard Problem

We adopt as our standard of reference problem (1) with $\rho = \text{const}$ and $g = \text{const}$ on an infinite-sheeted Riemann surface $r \geq \rho$, $-\infty < \varphi < \infty$, where r and φ are polar coordinates. We denote the corresponding Green function by $\Gamma(r_0, \varphi_0; r, \varphi; \kappa)$. This function (see [3]) has the form

$$\Gamma(r_0, \varphi_0; r, \varphi; \kappa) = \frac{i}{8} \int_C e^{i\nu|\varphi - \varphi_0|} H_\nu^{(1)}(\kappa r_>) \left[H_\nu^{(2)}(\kappa r_<) - \frac{H_\nu^{(2)\prime}(\kappa\rho) + ig H_\nu^{(2)}(\kappa\rho)}{H_\nu^{(1)\prime}(\kappa\rho) + ig H_\nu^{(1)}(\kappa\rho)} H_\nu^{(1)}(\kappa r_<) \right] d\nu, \tag{1.1}$$

where $r_{\gtrless}$ is the larger (smaller) of the numbers r and r_0 and the contour C may be regarded in the ν plane as consisting of the two half-lines $(\infty e^{2\pi i/3}, \kappa\rho)$ and $(\kappa\rho, +\infty)$.

We let $\gamma = \kappa\rho$ and $q = ig$. The denominator of the integrand in (1.1) has for any $\arg q$ a countable set of zeros $\nu_1, \nu_2, \ldots$ in the ν plane as $\gamma \to \infty$, close to the analogous zeros of the function $H_\nu^{(1)}(\gamma)$ on the line (A.4) (see Appendix). For certain complex values of q this denominator has, besides those indicated, a singular zero $\nu = \nu_0$ in the domain between the branch cut and the line (A.4) running to infinity in the first quadrant, i.e., for

$$0 \leq \arg \psi(\nu) < \frac{\pi}{2}, \tag{1.2}$$

where

$$\psi(\nu) = \nu \ln \frac{\nu + \sqrt{\nu^2 - \gamma^2}}{\gamma} - \sqrt{\nu^2 - \gamma^2}; \tag{1.3}$$

when $\nu > \gamma$ the radical and logarithm are assumed to be arithmetic. Outside the domain (1.2) the denominator does not have zeros for any q as $\gamma \to \infty$.

We wish to analyze the narrower domain

$$0 \leq \arg \psi(\nu) \leq \frac{\pi}{2} - \varepsilon, \tag{1.4}$$

where $\varepsilon > 0$ is a constant. In this domain $|s_\nu^{(2)}(\gamma)| \gg |s_\nu^{(1)}(\gamma)|$ (see Appendix), and the equation for ν_0 assumes the following form on the basis of Eqs. (A.1), (A.2), and (A.3):

$$-\frac{\sqrt{\nu^2 - \gamma^2}}{\gamma} \left[1 - \tilde{\Omega} + O\left(\frac{1}{\gamma^2}\right)\right] + q \left[1 - \Omega + O\left(\frac{1}{\gamma^2}\right)\right] + O(e^{-c\gamma}) = 0, \ \mathrm{Re}\, c > 0.$$

Hence, considering the explicit forms of Ω and $\tilde{\Omega}$, we find

$$\nu_0 = \gamma\sqrt{1 + q^2} + \frac{1}{2q\sqrt{1 + q^2}} + O(\gamma^{-1}). \tag{1.5}$$

Equation (1.5) holds only as long as the root ν_0 is in the domain (1.4), i.e., for values of q satisfying the inequality

$$0 \leq \arg \left[\sqrt{1 + q^2} \ln \left(q + \sqrt{1 + q^2}\right) - q\right] \leq \frac{\pi}{2} - \varepsilon. \tag{1.6}$$

As $\arg q$ increases (for fixed $|q|$) the root ν_0 approaches the line of zeros (A.4) and is subject to a more complicated condition than (1.5). It can be verified that with a further increase in $\arg q$ the root ν_0 moves along the line (A.4), staying at all times to the right of it, and as $\arg q \to \frac{\pi}{2}$ it goes to infinity in the first quadrant.

Applying the theorem of residues to the integral (1.1), we find

$$\Gamma(r_0, \varphi_0; r; \varphi; \kappa) = \sum_{p=1}^{\infty} \Gamma_p(r_0, \varphi_0; r, \varphi; \kappa) + \Gamma_0(r_0, \varphi_0; r, \varphi; \kappa). \tag{1.7}$$

The terms $\Gamma_p(r_0, \varphi_0; r, \varphi; \kappa)$, $p = 1, 2, \ldots$, are induced by the residues at $\nu = \nu_p$, $p = 1, 2, \ldots$. It is well known that they have the sense of slip waves traveling continually toward the circle $r = \rho$ from the source $\mathcal{M}_0$ along tangents moving around the circle with an exponential decay proportional to $\kappa^{1/3}$ and slipping along the tangents to the point of observation $\mathcal{M}$. The term $\Gamma(r_0, \varphi_0; r, \varphi; \kappa)$ is induced by the residue at $\nu = \nu_0$ and may be written in the form

$$\Gamma_0(r_0, \varphi_0; r, \varphi; \kappa) = \frac{1}{i} u^{\pm}(r, \varphi, \kappa) u^{\mp}(r_0, \varphi_0, \kappa), \quad \varphi \gtrless \varphi_0. \tag{1.8}$$

Here

$$u^{\pm}(r, \varphi, \kappa) = \frac{\sqrt{\pi}}{2} \cdot e^{i\frac{\pi}{4}} \left\{ \frac{H_{\nu_0}^{(2)\prime}(\kappa\rho) + q H_{\nu_0}^{(2)}(\kappa\rho)}{\frac{\partial}{\partial\nu}\left[H_{\nu}^{(1)\prime}(\kappa\rho) + q H_{\nu}^{(1)}(\kappa\rho)\right]_{\nu_0}} \right\}^{1/2} H_{\nu_0}^{(1)}(\kappa r) e^{\pm i\nu_0\varphi}. \tag{1.9}$$

Let

$$\kappa(r - \rho) = y, \qquad \kappa(r_0 - \rho) = y_0. \tag{1.10}$$

We shall assume that the parameter q satisfies condition (1.6) and that the observer and source are close enough to the circle $r = \rho$ that the quantities y and y_0 are finite. We then obtain for $u^{\pm}(r, \varphi, \kappa)$ on the basis of Eq. (1.5) and the equations of the Appendix:

$$u^{\pm}(r, \varphi, \kappa) = \frac{q^{1/2}}{(1+q^2)^{1/4}} \exp\left[\pm i\kappa\rho\sqrt{1+q^2}\,\varphi \pm \frac{i\varphi}{2q\sqrt{1+q^2}} - qy\right]\left[1 + O\left(\frac{1}{\kappa\rho}\right)\right]. \tag{1.11}$$

Equations (1.8) and (1.11) determine the asymptotic expression for $\Gamma_0(r_0, \varphi_0, r, \varphi; \kappa)$ when y and y_0 are finite. For $\mathrm{Re}\, q > 0$, obviously, this expression has the sense of a surface wave traveling along the boundary $r = \rho$ with a certain phase velocity v_φ. When $q > 0$,

$$v_\varphi = \frac{c}{\sqrt{1+q^2}}\left[1 - \frac{1}{2\kappa\rho(1+q^2)} + O\left(\frac{1}{\kappa^2}\right)\right], \tag{1.12}$$

where c is the constant wave propagation velocity in the medium surrounding the contour $r = \rho$.

When $q > 0$, the surface wave (1.11) travels along the boundary without damping. A small positive value of $\arg q$ corresponds to small surface-wave damping. As $\arg q$ increases, the damping does likewise, being characterized by the factor

$$\exp\left[i\kappa\rho\,|\varphi - \varphi_0|\, \mathrm{Im}\sqrt{1+q^2}\right], \tag{1.13}$$

which enters into the expression for Γ_0. The factor (1.13) shows that when the root ν_0 approaches the line (A.4) and $|\varphi - \varphi_0| > 0$ the quantity Γ_0 becomes exponentially small, i.e., the surface wave ceases to exist. Consequently, it is only meaningful to speak of a surface wave as long as $q > 0$ and $\arg q$ has small positive values, and the description of the position of the root ν_0 may be limited to Eq. (1.5).

The factor $\exp\left[i|\varphi - \varphi_0|\left(2q\sqrt{1+q^2}\right)^{-1}\right]$ is missing from the expression for Γ_0 in [4]. This is a result of the insufficient accuracy of the equation for ν_0, which does not include the term $\left(2q\sqrt{1+q^2}\right)^{-1}$ [see

(1.5)]; this term affects the principal term of the asymptotic representation of Γ_0 and is related to the inclusion of the correction terms Ω and $\tilde{\Omega}$ in Eqs. (A.2) and (A.3).

Next we consider the Green function $G(r_0, \varphi_0; r, \varphi; \kappa)$ for the exterior of the circle, not on an infinite-sheeted Riemann surface, but on an ordinary plane. This function can be formulated (see [5]) from the relation

$$G(r_0, \varphi_0; r, \varphi; \kappa) = \sum_{j=-\infty}^{\infty} \Gamma(r_0, \varphi_0; r, \varphi + 2\pi j; \kappa). \tag{1.14}$$

By virtue of (1.8) and (1.9)

$$\Gamma_0(r_0, \varphi_0; r, \varphi + 2\pi j; \kappa) = \frac{1}{i} u^{\pm}(r, \varphi, \kappa) u^{\mp}(r_0, \varphi_0, \kappa) e^{\pm 2\pi i j \nu_0}, \tag{1.15}$$

$$\varphi \gtrless \varphi_0, \quad j = 0, \pm 1, \ldots .$$

Therefore,

$$G_0(r_0, \varphi_0; r, \varphi; \kappa) = \sum_{j=-\infty}^{\infty} \Gamma_0(r_0, \varphi_0; r, \varphi + 2\pi j; \kappa) = \frac{1}{i} \Big[u^{\mp}(r_0, \varphi_0, \kappa) u^{\pm}(r, \varphi, \kappa) +$$

$$+ \frac{u^{-}(r_0, \varphi_0, \kappa) u^{+}(r, \varphi, \kappa) + u^{+}(r_0, \varphi_0, \kappa) u^{-}(r, \varphi, \kappa)}{1 - e^{2\pi i \nu_0}} e^{2\pi i \nu_0} \Big], \quad \varphi \gtrless \varphi_0. \tag{1.16}$$

This equation implies that the function $G(r_0, \varphi_0; r, \varphi; \kappa)$, due to the complexity of q, has on the κ plane additional poles defined by the equation

$$\nu_0(\kappa) = \ell, \tag{1.17}$$

in which ℓ is an integer. Substituting Eq. (1.5) into (1.17), we obtain the following explicit asymptotic equation for the additional series of poles:

$$\kappa_\ell = \frac{\ell}{\rho\sqrt{1+q^2}} - \frac{1}{2\rho(1+q^2)} + O\left(\frac{1}{\ell}\right), \quad \ell \gg 1. \tag{1.18}$$

When $q > 0$, these additional poles are real. As $\arg q$ increases they move into the lower half-plane of κ.

§ 2. Surface Waves in the General Case

In order to ascertain the surface wave in the case of variable and sufficiently smooth functions $\rho(s)$ and $q(s) = -i\,q(s)$ satisfying conditions (2) and (3), we resort to a slightly modified ray method. We write the homogeneous Helmholtz equation and impedance boundary condition in coordinates s and $\nu = n\kappa$:

$$\left(1 + \frac{\nu}{\kappa\rho(s)}\right)^2 \kappa^2 u_{\nu\nu} + \frac{\kappa}{\rho(s)} u_\nu + \left(1 + \frac{\nu}{\kappa\rho(s)}\right)^{-1} u_{ss} + \frac{\kappa\nu\rho'(s)}{(\kappa\rho(s) + \nu)^2} u_s + \left(\kappa^2 + \frac{\kappa\nu}{\rho(s)}\right) u = 0, \tag{2.1}$$

$$u_\nu + q(s) u \big|_{\nu=0} = 0 \tag{2.2}$$

The radius of curvature $\rho(s)$ is assumed to be positive when the contour S is convex toward the observation point $\mathcal{M}$, and negative when S is concave relative to $\mathcal{M}$. In the case of a constant radius of curvature the coordinate ν reverts to the coordinate y of (1.10). The introduction of the coordinates s and ν (ν is considered to be finite when $\kappa \to \infty$) in place of the conventional ray coordinates is extremely helpful in the analysis of surface waves whose intensity decays along n with a coefficient proportional to κ. We seek the solution of Eq. (2.1) under condition (2.2) in the form

$$u(\mathcal{M},\kappa)=e^{i\kappa\tau(s)-\nu\varphi(s)}\left[\sum_{m=0}^{N-1}\frac{a_m(s,\nu)}{(i\kappa)^m}+O\left(\kappa^{-N}\right)\right],\quad \mathcal{M}=(s,\nu),\quad N=1,2,\ldots. \tag{2.3}$$

We assume that the functions we are after, $a_m(s,\nu)$, are polynomials in ν with unknown coefficients depending on s.* The customary procedure of substituting (2.3) into (2.1) and (2.2), setting the coefficients of each power of κ equal to zero, and determining the degree of the polynomials $a_m(s,\nu)$ yields a sequence for the determination of the functions $\tau(s), \varphi(s), a_0(s,\nu), a_1(s,\nu), \ldots$.

As an example we consider the first step in these calculations, in which the functions $\tau(s)$, $\varphi(s)$ and the degree of the polynomial a_0 are determined. The "leading" equation, which evolves from (2.1) when the terms with κ^2 are set equal to zero, has the form

$$a_0(\varphi^2-\tau'^2+1)-2\varphi\frac{\partial a_0}{\partial\nu}+\frac{\partial^2 a_0}{\partial\nu^2}=0. \tag{2.4}$$

We denote by l the degree of the polynomial $a_0(s,\nu)$:

$$a_0(s,\nu)=a_{0l}(s)\nu^l+a_{0,l-1}(s)\nu^{l-1}+\ldots+a_{00}(s),\quad a_{0l}(s)\not\equiv 0.$$

We assume that $l\geqslant 1$. Then Eq. (2.4) may be written

$$\nu^l(\varphi^2-\tau'^2+1)a_{0l}(s)+\nu^{l-1}\left[(\varphi^2-\tau'^2+1)a_{0,l-1}(s)-2\varphi a_{0l}(s)\right]+\ldots=0.$$

We consider the coefficient of ν^l. Inasmuch as $a_{0l}(s)\not\equiv 0$, it follows that

$$\varphi^2-\tau'^2+1=0. \tag{2.5}$$

Regressing to the coefficient of ν^{l-1}, we infer that the equation $\varphi=0$ must hold. This is impossible, however, because in this event the solution (2.3) would no longer have surface-wave character. The only admissible assumption is that $l=0$. Then Eq. (2.5) remains valid. We now consider the boundary conditions (2.2). Assembling terms of order κ, we obtain

$$\varphi(s)=q(s).$$

Hence, according to (2.5), we deduce

$$\tau(s)=\pm\int_0^s\sqrt{1+q^2(s)}\,ds. \tag{2.6}$$

We denote the solutions of problem (2.1)-(2.2) corresponding to the two sign choices in Eq. (2.6) by $u^{\pm}(\mathcal{M},\kappa)$.

The next step, i.e., setting terms with κ to the first power in Eq. (2.1) and with κ to the zeroth power in (2.2) equal to zero, determines the degree of the polynomial a_1[a_1 turns out to be a second-degree polynomial: $a_1=a_{12}(s)\nu^2+a_{11}(s)\nu+a_{10}(s)$] as well as the functions $a_0(s)$, $a_{12}(s)$, $a_{11}(s)$.

We now give (up to arbitrary constant multipliers) the final result of the calculations:

*A similar assumption has been tendered in [6] with respect to another problem. In the special case of the problem described in the preceding section, the function $a_m(s,\nu)$ does not actually have a polynomial character.

$$u^{\pm}(\mathcal{M},\kappa)=\frac{q^{1/2}(s)}{\sqrt[4]{1+q^2(s)}}\exp\left[\pm i\kappa\int_0^s\sqrt{1+q^2(s)}\,ds-\nu q(s)\pm\frac{i}{2}\int_0^s\frac{ds}{\rho(s)q(s)\sqrt{1+q^2(s)}}\right]\left[1\pm\frac{b^{\pm}(s,\nu)}{i\kappa}+O\left(\frac{1}{\kappa^2}\right)\right],$$

$$b^{\pm}(s,\nu)=\pm\frac{i\sqrt{1+q^2(s)}}{2q(s)}\left(\frac{\sqrt{1+q^2(s)}}{\rho(s)}\mp iq'(s)\right)\nu^2+b_0^{\pm}(s), \tag{2.7}$$

where $b_0^{\pm}(s)$ represents rather intricate expressions, which we shall not write out explicitly. The $u^{\pm}(\mathcal{M},\kappa)$ have the sense of surface waves traveling along the boundary S in the directions of increasing and decreasing s. These functions are concentrated for

$$\nu\leq \text{const}=\underset{\kappa\to\infty}{O(1)} \tag{2.8}$$

in the sense that they become exponentially small outside the layer (2.8). In order to characterize the surface-wave damping with distance from S we introduce the scale ℓ, i.e., the distance over which the wave amplitude is diminished by a factor of e. We conclude on the basis of (2.7) that

$$\ell=(\kappa\,\mathrm{Re}\,q)^{-1}. \tag{2.9}$$

The phase velocity of the surface waves (2.7) for $q>0$ is equal to

$$v_{\varphi}(s)=\frac{c}{\sqrt{1+q^2(s)}}\left[1-\frac{1}{2\kappa\rho(s)q(s)\,[1+q^2(s)]}+O(\kappa^{-2})\right], \tag{2.10}$$

where c is the propagation velocity. In the case of constant ρ and q, Eqs. (2.7) and (2.10) revert to Eqs. (1.11) and (1.12). It is reasonable, therefore, to consider the functions $u^{\pm}(\mathcal{M},\kappa)$ as a generalization of the functions $u^{\pm}(\rho,\varphi,\kappa)$ to the case of variable $\rho(s)$ and $q(s)$.

All the calculations of the present section are valid regardless of the sign of the curvature $\rho^{-1}(s)$, which can revert to zero. When $\rho^{-1}(s)<0$, all that is required is that the point $\mathcal{M}$ be close enough to S that its position may be uniquely characterized by the coordinates ν and s. In the principal surface-wave zone (2.8) the aforementioned requirement is always met for sufficiently large κ and smooth S.

The influence of the sign of the curvature $\rho^{-1}(s)$ of S is felt, according to (2.10), in the sign of the correction to the expression $c\,[1+q^2(s)]^{-1/2}$ for the phase velocity. If $\rho^{-1}(s)>0$ (contour S convex relative to the point $\mathcal{M}$), anomalous dispersion takes place. If $\rho^{-1}(s)<0$ (contour S concave relative to $\mathcal{M}$), the dispersion is normal. When $\rho^{-1}(s)=0$, the sign of the dispersion depends on higher terms of the asymptotic representation.

§ 3. Surface-Wave Excitation in the General Case

Having formulated (for $\kappa\to\infty$) the generalization of the expressions for $u^{\pm}(r,\varphi,\kappa)$ to the case of variable functions $\rho(s)$ and $q(s)$, we are logically motivated to seek an analogous generalization for $\Gamma_0(r_0,\varphi_0;r,\varphi;\kappa)$. We do so by referring to Eq. (1.8).

We begin with the case of an infinite contour S. The effect of surface-wave excitation at an observation point $\mathcal{M}(s,\nu)$ by a point source at $\mathcal{M}_0(s_0,\nu_0)$ may be sought in the following form in the case of sufficiently smooth $\rho(s)$ and $q(s)$ [subject to conditions (2) and (3)]:

$$\Gamma_0(\mathcal{M}_0,\mathcal{M},\kappa)=A u^{\mp}(\mathcal{M}_0,\kappa)u^{\pm}(\mathcal{M},\kappa),\quad s\gtrless s_0, \tag{3.1}$$

where the factor A is to be determined.

This factor is independent of either the coordinates of the point $\mathcal{M}$ or the coordinates of $\mathcal{M}_0$. Under this condition Eq. (3.1) guarantees that the function $\Gamma_0(\mathcal{M}_0,\mathcal{M},\kappa)$ will satisfy Eq. (1) outside the source, as well as the impedance boundary condition. Equation (3.1) guarantees symmetry of the component Γ_0 of the Green function (4).*

Inasmuch as the functions $u^{\pm}(\mathcal{M},\kappa)$ and $\Gamma_0(\mathcal{M}_0,\mathcal{M},\kappa)$ are dimensionless [see (1) and (2.7)], the factor $\mathcal{A}$ is dimensionless.

We have recourse to arguments (called the localization principle and developed in connection with the high-frequency investigation of special problems) in support of the fact that the dimensionless factor $\mathcal{A}$, being independent of $\mathcal{M}_0$ and $\mathcal{M}$, is also independent of either κ or the configuration of the contour S, i.e., is a universal constant.† According to the localization principle the high-frequency asymptotic behavior of the Green function is wholly determined by the properties of the path of the corresponding ray or rays or, in the case of $\Gamma_0(\mathcal{M}_0,\mathcal{M},\kappa)_-$, by the properties of the contour S between the projections P_0 and P of the points $\mathcal{M}_0$ and $\mathcal{M}$ onto that contour. The implication is that the factor $\mathcal{A}$ cannot depend on the properties of S as a whole, but can depend on the segment $\breve{P_0P}$ of the contour. However, neither can $\mathcal{A}$ depend on the arc $\breve{P_0P}$, because the latter varies with the positions of $\mathcal{M}_0$ and $\mathcal{M}$, and $\mathcal{A}$ is independent of the positions of these points. But then $\mathcal{A}$ cannot depend on κ, because if it is independent of S one cannot form a dimensionless parameter involving κ.

We assume, therefore, that the factor $\mathcal{A}$ is independent of both κ and the form of S. It can then be determined from the exact solution of a particular problem, for example, on the basis of the exact solution of (1.1). Since the functions $u^{\pm}(\mathcal{M},\kappa)$ revert to the functions $u^{\pm}(r,\varphi,\kappa)$ when $\rho=\text{const}>0$ and $q=\text{const}$, it follows that‡

$$\mathcal{A}=\frac{1}{i}. \tag{3.2}$$

Hence,

$$\Gamma_0(\mathcal{M}_0,\mathcal{M},\kappa)=\frac{\sqrt{q(s_0)q(s)}}{i\sqrt[4]{1+q^2(s_0)}\,\sqrt[4]{1+q^2(s)}}\exp\left[i\kappa\int_{s_>}^{s_>}\sqrt{1+q^2(s)}\,ds-\nu q(s)-\nu_0 q(s_0)+\frac{i}{2}\int_{s_<}^{s_>}\frac{ds}{q(s)\rho(s)\sqrt{1+q^2(s)}}\right]\left[1+O\left(\frac{1}{\kappa}\right)\right], \tag{3.3}$$

where $s_>$ and $s_<$ represent the larger and the smaller, respectively, of the quantities s_0 and s.

Inasmuch as the asymptotic expressions for $u^{\pm}(\mathcal{M},\kappa)$ used in the formulation of Eq. (3.3) are equally applicable to the cases of positive, negative, or zero curvature $\rho^{-1}(s)$, it is reasonable to assume that Eq. (3.3) is also valid in all of these cases.

Now let S be a closed contour of length L. We put it in correspondence with an infinite-sheeted Riemann surface, $n\geq 0,\ -\infty<s<\infty$, analogous to the Riemann surface of §1. Let $\Gamma_0(\mathcal{M}_0,\mathcal{M},\kappa)$ be the part of the Green function on this surface which is responsible for surface-wave excitation. In order to find an L-periodic function $G_0(\mathcal{M}_0,\mathcal{M},\kappa)$ describing the excitation of a surface wave in the case of a closed contour, we need to sum the expressions already found for $\Gamma_0(\mathcal{M}_0,\mathcal{M}^j,\kappa)$ over all points of observation $\mathcal{M}^j=(s+jL,\nu)$, $j=0,\pm 1,\ldots$, situated "one under the other" on the Riemann surface. We obtain

$$G_0(\mathcal{M}_0,\mathcal{M},\kappa)=\frac{1}{i}\left[1-e^{i\Phi(\kappa)}\right]^{-1}\left[u^{\mp}(\mathcal{M}_0,\kappa)u^{\pm}(\mathcal{M},\kappa)+e^{i\Phi(\kappa)}u^{\pm}(\mathcal{M}_0,\kappa)u^{\mp}(\mathcal{M},\kappa)\right],\quad s\gtrless s_0, \tag{3.4}$$

$$\Phi(\kappa)=\kappa\int_0^L\sqrt{1+q^2(s)}\,ds+\frac{1}{2}\int_0^L\frac{ds}{\rho(s)q(s)\sqrt{1+q^2(s)}}+O(\kappa^{-1}). \tag{3.5}$$

*The other component of the Green function, $\tilde{\Gamma}(\mathcal{M}_0,\mathcal{M},\kappa)$, has been formulated for $\kappa\to\infty$ in [7] and also has symmetry.

†The ensuing considerations stem from [8]. They have also been profoundly influenced by [9]. Similar considerations are given in [10] for the other part $\tilde{\Gamma}(\mathcal{M}_0,\mathcal{M},\kappa)$ of the Green function.

‡The same expression for $\mathcal{A}$, i.e., Eq. (3.2), can be obtained on the basis of a comparison with problem (1.1) for $\rho^{-1}(s)=0$ (S is a straight line) and $q=\text{const}$.

Equations (3.4) and (3.5) show that the function $G_0(\mathcal{M}_0, \mathcal{M}, \kappa)$ and, together with it, the total Green function $G(\mathcal{M}_0, \mathcal{M}, \kappa)$ have the following additional poles in the κ plane for a closed contour S [subject to conditions (2) and (3)]:

$$\kappa = \kappa_l \equiv \left(\int_0^L \sqrt{1+q^2(s)}\, ds \right)^{-1} \left[2\pi l - \frac{1}{2} \int_0^L \frac{ds}{\rho(s) q(s) \sqrt{1+q^2(s)}} + O(l^{-1}) \right], \quad l = l_0, l_0+1, \ldots, \quad l_0 >> 1. \tag{3.6}$$

For $\rho(s) = \text{const}$ and $q(s) = \text{const}$ these poles revert to the poles (1.18).

§ 4. Frequency-Dependent Impedance Case

It is necessary in the solution of practical problems to investigate the more general [than in (1)] impedance condition in which

$$q(s) = -iq(s, \kappa) = -i \sum_{m=0}^{\infty} \frac{q_m(s)}{(i\kappa)^m}. \tag{4.1}$$

Among the problems that reduce to an impedance boundary condition with a function $q(s)$ of the form (4.1) are the problems of Rayleigh and Stoneley waves (see [11]), waves near the surface of a conductor with a dielectric coating, and waves near corrugated or interdigital surfaces.

The entire scheme of the calculations from § 2 and § 3 carry over to the case (4.1), only the final equations being modified. Instead of (2.7) and (3.3) we obtain

$$u^{\pm}(\mathcal{M}, \kappa) = \frac{q_0^{1/2}(s)}{\sqrt[4]{1+q_0^2(s)}} \exp\left[\pm i\kappa \int_0^s \sqrt{1+q_0^2(s)}\, ds - \nu q_0(s) \pm \frac{i}{2} \int_0^s \frac{ds}{\rho(s) q_0(s) \sqrt{1+q_0^2(s)}} \pm \int_0^s \frac{q_0(s) q_1(s)}{\sqrt{1+q_0^2(s)}} ds \right] \left[1 + O(\kappa^{-1}) \right],$$

$$\Gamma_0(\mathcal{M}_0, \mathcal{M}; \kappa) = \frac{\sqrt{q_0(s_0) q_0(s)}}{i \sqrt[4]{1+q_0^2(s_0)} \sqrt{1+q_0^2(s)}} \exp\left[i\kappa \int_{s_<}^{s_>} \sqrt{1+q_0^2(s)}\, ds - \nu q_0(s) - \right.$$

$$\left. - \nu_0 q_0(s_0) + \frac{i}{2} \int_{s_<}^{s_>} \frac{ds}{\rho(s) q_0(s) \sqrt{1+q_0^2(s)}} + \int_{s_<}^{s_>} \frac{q_0(s) q_1(s)}{\sqrt{1+q_0^2(s)}} ds \right] \left[1 + O(\kappa^{-1}) \right].$$

Equations (4.2) and (4.3) show that the impedance correction $q_1(s)$ affects even the principal terms of the asymptotic representations of $u^{\pm}(\mathcal{M}, \kappa)$ and $\Gamma_0(\mathcal{M}_0, \mathcal{M}, \kappa)$.

Appendix

In order to facilitate the discussion of § 1 we give the Debye asymptotic relations (see [12]) for the Hankel functions $H_\nu^{(1)}(\gamma)$ and $H_\nu^{(2)}(\gamma)$ and their derivatives for $\gamma > 0$ in the first quadrant of the complex plane of ν, from which we cut out, adjacent to the origin, a curvilinear triangle OAB and semicircle of radius $O(\gamma^{1/3})$ with center at $\nu = \gamma$ (see Fig. 1). We treat γ and $|\nu|$ as large parameters. The segment AB marked in the figure by the thin solid line is described in the ν plane by the equation

$$\arg \Psi(\nu) = \pi$$

[see (1.3)]. Branch cuts are made in the ν plane for $\nu > \gamma$ and $\nu < -\gamma$, and the radical and logarithm are assumed to be arithmetic on the upper border of the cut $\nu > \gamma$.

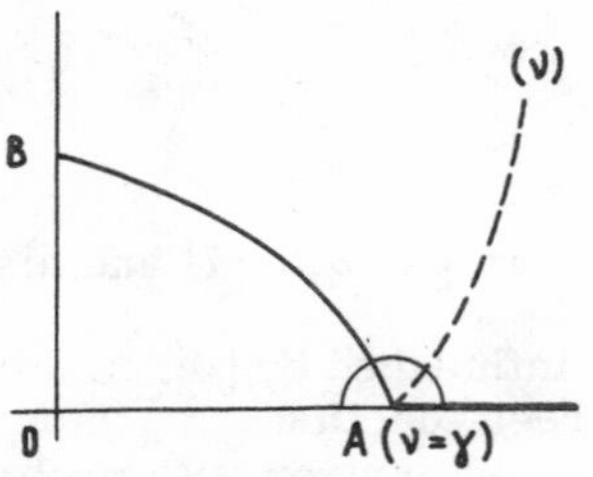

Fig. 1.

The following equations hold in the designated part of the first quadrant:

$$H_\nu^{(1)}(\gamma) = S_\nu^{(1)}(\gamma) - S_\nu^{(2)}(\gamma), \quad H_\nu^{(2)}(\gamma) = S_\nu^{(2)}(\gamma), \tag{A.1}$$

in which $S_\nu^{(1)}(\gamma)$ and $S_\nu^{(2)}(\gamma)$ have the following asymptotic representations:

$$\left.\begin{aligned} S_\nu^{(1)}(\gamma) &= \sqrt{\tfrac{2}{\pi}}\,(\nu^2-\gamma^2)^{-1/4} e^{-\psi(\nu)}\left[1+\Omega+O(\gamma^{-2})\right],\\ S_\nu^{(2)}(\gamma) &= i\sqrt{\tfrac{2}{\pi}}\,(\nu^2-\gamma^2)^{-1/4} e^{\psi(\nu)}\left[1-\Omega+O(\gamma^{-2})\right],\\ \Omega &= \frac{1}{8}\left(\frac{1}{\sqrt{\nu^2-\gamma^2}} - \frac{5}{3}\,\frac{\nu^2}{(\nu^2-\gamma^2)^{3/2}}\right). \end{aligned}\right| \tag{A.2}$$

We also give expressions for the derivatives $S_\nu^{(1)\prime}(\gamma)$ and $S_\nu^{(2)\prime}(\gamma)$, in terms of which, according to (A.1), the derivatives of the Hankel functions are expressed:

$$\left.\begin{aligned} S_\nu^{(1)\prime}(\gamma) &= \sqrt{\tfrac{2}{\pi}}\,(\nu^2-\gamma^2)^{1/4}\gamma^{-1} e^{-\psi(\nu)}\left[1+\tilde{\Omega}+O(\gamma^{-2})\right],\\ S_\nu^{(2)\prime}(\gamma) &= -i\sqrt{\tfrac{2}{\pi}}\,(\nu^2-\gamma^2)^{1/4}\gamma^{-1} e^{\psi(\nu)}\left[1-\tilde{\Omega}+O(\gamma^{-2})\right],\\ \tilde{\Omega} &= -\frac{1}{8}\left(\frac{3}{\sqrt{\nu^2-\gamma^2}} - \frac{7}{3}\,\frac{\nu^2}{(\nu^2-\gamma^2)^{3/2}}\right). \end{aligned}\right| \tag{A.3}$$

It follows from Eqs. (A.1), (A.2), and (A.3) that the functions $H_\nu^{(1)}(\gamma)$ and $H_\nu^{(1)\prime}(\gamma)$ have in the indicated domain a countable set of zeros on the line

$$\arg \psi(\nu) = \frac{\pi}{2}, \tag{A.4}$$

which is marked by a broken line in Fig. 1.

LITERATURE CITED

1. Miller, M. A., and Talanov, V. I., Application of the surface impedance concept in the theory of electromagnetic surface waves (review), Izv. VUZov, Radiofizika, 4(5):795-830 (1961).
2. Grimshaw, R., Propagation of surface waves at high frequencies, J. Inst. Math. Appl., 4:174-193 (1968).
3. Vainshtein, L. A., and Malyuzhinets, G. D., Transverse diffusion in diffraction by an impedance cylinder of large radius (part II), Radiotekh. i Elektron., 6(9):1489-1495 (1961).
4. Keller, J. B., and Karal, F. C., Surface-wave excitation and propagation, J. Appl. Phys., 31(6): 1039-1046 (1960).
5. Friedlander, F. G., Sound Pulses, Cambridge University Press (1958).
6. Lazutkin, V. F., Asymptotic representation of the eigenfunctions, concentrated near the boundary of a region, of the Laplace operator, Zh. Vychis. Matem. i Matem. Fiz., 7(6):1237-1249 (1967).
7. Molotkov, Green function for the problem of diffraction by a convex cylinder with variable impedance, Trudy Matem. Inst. Steklov, 95:119-131 (1968).
8. Keller, J. B., Diffraction by a convex cylinder, Trans. IRE Antennas and Propagation, AP-4(3): 312-321 (1956).
9. Buldyrev, V. S., Short-wave interference in the problem of diffraction by an inhomogeneous cylinder of nonuniform cross section, Izv. VUZov, Radiofizika, 10(5):699-711 (1967).
10. Molotkov, I. A., The Field of a point source located outside a convex curve, Topics in Mathematical Physics, Vol. 4, Consultants Bureau, New York (1971), pp. 77-100.

11. Molotkov, Excitation of Rayleigh and Stoneley waves, this volume, pp. 93-101.
12. Petrashen', G. I., Makarov, G. I., and Smirnov, N. S., On the asymptotic representations of cylindrical functions, Uch. Zap. Leningrad. Univ. (LGU), No. 170, pp. 7-95 (1953).

EXCITATION OF RAYLEIGH AND STONELEY WAVES

I. A. Molotkov

The present article is concerned with the point-source excitation of Rayleigh and Stoneley waves on the boundary of an elastic medium with a liquid or gas. We solve the problem in the short-wave approximation, using the same method as in [1]. At the end of the article we also investigate the limiting case in which the elastic medium bounds a vacuum and undamped Rayleigh waves are generated.

§ 1. Surface Waves near the Boundary of an Elastic Medium with a Liquid or Gas

Let a plane curve S described by a realistic equation $\rho=\rho(s)$ represent the boundary between a homogeneous elastic medium and a homogeneous liquid or gas, let Φ and Ψ denote the longitudinal and transverse potentials in the elastic medium (domain $n \geq 0$, where n is the length of the normal), and let Φ_1 be the displacement potential in the liquid or gas (domain $n \leq 0$). We consider the plane problem of solving the equations

$$(\Delta+k^2)\Phi(M_0, M, k) = -\delta(M-M_0), \quad k=\frac{\omega}{a},$$
$$(\Delta+\varkappa^2)\Psi(M_0, M, \varkappa) = 0, \quad \varkappa=\frac{\omega}{b} \tag{1.1}$$

(for $n>0$) and the equation

$$(\Delta+k_1^2)\Phi_1(M_0, M, k_1)=0, \quad k_1=\frac{\omega}{a_1} \tag{1.2}$$

(for $n<0$) under the boundary conditions

$$2\Phi_{sn}-2\Psi_{nn}-\varkappa^2\Psi\big|_S=0,$$
$$2(\Phi_{nn}+\Psi_{sn})-\frac{a^2-2b^2}{b^2}k^2\Phi+\rho_0\frac{a^2}{b^2}k^2\Phi_1\Big|_S=0, \tag{1.3}$$
$$\Phi_n+\Psi_s-\Phi_{1n}\big|_S=0$$

and the absorption limit conditions. The first two boundary conditions express the continuity of the normal stresses at S, while the last boundary condition describes the continuity of the normal component of the displacement vector at S. The following notation is used in Eqs. (1.1), (1.2), and (1.3): M_0 and M are the source and observation points, a and b are the propagation velocities of longitudinal and transverse waves in the elastic medium, a_1 is the propagation velocity in the liquid (gas), ρ_0 is the ratio of the liquid (gas) density to the elastic medium density, s is the arc length on S, and ω is the frequency. The boundary S is assumed to be sufficiently smooth and of infinite extent. We consider the curvature $\rho^{-1}(s)$ to be positive at points where S is concave relative to the elastic medium. In the case of a boundary of finite length the solution can be obtained by summing certain solutions for the infinite boundary case, as in the transition from Γ_0 to G_0 in [1].

It is required to segregate (as $\omega \to \infty$) from the functions $\Phi(M_0, M, \kappa)$, $\Psi(M_0, M, \varkappa)$, and $\Phi_1(M_0, M, \kappa_1)$ the terms corresponding to the excitation of surface waves by a point source.

We begin with the formulation of the solutions $\varphi(M, \kappa)$, $\psi(M, \varkappa)$, and $\varphi_1(M, \kappa_1)$ of the homogeneous Helmholtz equations, such as will satisfy the boundary conditions (2.3) and have the character of surface waves in the elastic medium. We assume that the observation point M is located sufficiently close to the boundary S (for finite values of $\nu = \kappa n$, $\bar{\nu} = \varkappa n$, and $\nu_1 = \kappa_1 n$), since only for such positions of M will the surface-wave amplitudes not be small. The solutions we seek are best formulated in the form of expansions similar to (2.3) in [1]:

$$\varphi(M, \kappa) = e^{i\kappa\tau(s) - \alpha\nu}\left[c_0(s, \nu) + \frac{c_1(s, \nu)}{i\kappa} + \ldots\right], \quad \nu = \kappa n \geq 0,$$

$$\psi(M, \varkappa) = e^{i\kappa\tau(s) - \beta\bar{\nu}}\left[d_0(s, \bar{\nu}) + \frac{d_1(s, \bar{\nu})}{i\varkappa} + \ldots\right], \quad \bar{\nu} = \varkappa n \geq 0, \tag{1.4}$$

$$\varphi_1(M, \kappa_1) = e^{i\kappa\tau(s) + i\gamma|\nu_1|}\left[e_0(s, \nu_1) + \frac{e_1(s, \nu_1)}{i\kappa_1} + \ldots\right], \quad \nu_1 = \kappa_1 n \leq 0.$$

The functions $c_0(s, \nu), d_0(s, \bar{\nu}), e_0(s, \nu_1), c_1(s, \nu), d_1(s, \bar{\nu}), e_1(s, \nu_1), \ldots$ are assumed to be polynomials with respect to their second arguments.

The unknown functions in expansions (1.4) are determined in the usual way by substituting them into the homogeneous Helmholtz equations and boundary conditions (1.3), then setting the coefficients of each power of κ equal to zero. As in [1], it is easily verified that the degrees of the polynomials c_0, d_0, and e_0 are equal to zero, i.e., $c_0(s, \nu) = c_0(s)$, $d_0(s, \bar{\nu}) = d_0(s)$, $e_0(s, \nu_1) = e_0(s)$. For the determination of these three functions of s the boundary conditions (1.3) yield three homogeneous linear equations. Setting the determinant of the resulting system equal to zero, we obtain the familiar (see [2], p. 41) equation for the quantity $c = \frac{a}{\tau'}$, which in the case $\kappa \to \infty$ has the sense of the surface-wave propagation velocity:

$$a(2b^2 - c^2)^2 - 4b^3\sqrt{a^2 - c^2}\sqrt{b^2 - c^2} = -i\varrho_0 a_1 c^4 \frac{\sqrt{a^2 - c^2}}{\sqrt{c^2 - a_1^2}}. \tag{1.5}$$

The radical $\sqrt{c^2 - a_1^2}$ is positive when $c > a_1$. When $c < a_1$, it is required on the right-hand side of (1.5) to put

$$\sqrt{c^2 - a_1^2} = i\sqrt{a_1^2 - c^2}, \tag{1.6}$$

where $\sqrt{a_1^2 - c^2} > 0$.

The right-hand side of Eq. (1.5) involves as a small quantity the ratio of the acoustic stiffness of the two media in question:

$$\xi = \varrho_0 \frac{a_1}{b} \ll 1. \tag{1.7}$$

When $\xi \to 0$, Eq. (1.5) goes over to the Rayleigh equation

$$a(2b^2-c^2)^2-4b^3\sqrt{a^2-c^2}\sqrt{b^2-c^2}=0, \tag{1.8}$$

which has a single positive root c_0 (the Rayleigh propagation velocity). If $\xi \neq 0$, Eq. (1.5) has a complex root $c=c_R$. Under condition (1.7) this root has the form

$$c_R=c_0(1-i\varepsilon), \tag{1.9}$$

$$\varepsilon=\frac{\xi}{4}\frac{abc_0^2\sqrt{\frac{a^2-c_0^2}{c_0^2-a_1^2}}\,(2b^2-c_0^2)^2}{4b^6(a^2+b^2-2c_0^2)-a^2(2b^2-c_0^2)}+O(\xi^2). \tag{1.10}$$

Using Eq. (1.8), we can show that the principal term of the right-hand side of (1.10) is positive. This fact corresponds to the propagation of a Rayleigh wave along S with damping determined by the quantity ε and caused by the radiation of energy into the liquid (gas).

Equation (1.5) for the velocity c has another solution $c=c_S$, which is close to the velocity a_1 in the liquid:

$$c_S=a_1(1-\delta). \tag{1.11}$$

Assuming in (1.5) that

$$a_1^2 \ll b^2, \quad \delta \ll 1, \tag{1.12}$$

and taking (1.6) into account, we obtain

$$\delta=\frac{1}{8}\left(\varrho_0\frac{a_1^2}{b^2}\frac{a^2}{a^2-b^2}\right)^2\left[1+O\left(\frac{a_1^2}{b^2}\right)\right]>0. \tag{1.13}$$

Corresponding to the positive solution (1.11) is a second surface wave that can occur at the boundary of an elastic medium with a liquid (gas). This wave is called a Stoneley wave (see [3]).

The problem stated at the beginning of this section can now be particularized. It is required to segregate (as $\omega \to \infty$) from the solutions $\Phi(M_0, M, \kappa)$, $\Psi(M_0, M, \varkappa)$, and $\Phi_1(M_0, M, \kappa_1)$ of problem (1.1)-(1.3) (under the absorption limit conditions) the terms $\Phi_R, \Psi_R, \Phi_{1R}$ and $\Phi_S, \Psi_S, \Phi_{1S}$ describing the excitation of damped Rayleigh and Stoneley waves by a point source.

§ 2. Excitation of Rayleigh and Stoneley Waves

We now find the expressions for $\Phi_R, \Psi_R, \Phi_{1R}$ and $\Phi_S, \Psi_S, \Phi_{1S}$ describing the excitation of Rayleigh and Stoneley waves by a point source. We first give the result of formulating solutions $\varphi^{\pm}(M,\kappa)$, $\psi^{\pm}(M,\varkappa)$, and $\varphi_1^{\pm}(M,\kappa_1)$ of the form (1.4) of the homogeneous Helmholtz equations, subject to the boundary conditions (1.3). These solutions have the following form for $s>0$, up to a common arbitrary constant multiplier:

$$\varphi^{\pm}(M,\kappa)=\exp\left[\pm\frac{i\omega s}{c}-n\omega\frac{\sqrt{a^2-c^2}}{ac}\pm i\chi\int_0^s\frac{ds}{\rho(s)}\right]\left\{1+\left[c_{12}(s)\left(\frac{n\omega}{a}\right)^2+c_{11}(s)\frac{n\omega}{a}+c_{10}(s)\right]\frac{a}{i\omega}+O(\omega^{-2})\right\}, \tag{2.1}$$

$$\psi^{\pm}(M,\varkappa)=\mp i\Lambda\exp\left[\pm\frac{i\omega s}{c}-n\omega\frac{\sqrt{b^2-c^2}}{bc}\pm i\chi\int_0^s\frac{ds}{\rho(s)}\right]\left\{1+\left[d_{12}(s)\left(\frac{n\omega}{b}\right)^2+d_{11}(s)\frac{n\omega}{b}+d_{10}(s)\right]\frac{b}{i\omega}+O(\omega^{-2})\right\}, \tag{2.2}$$

$$\varphi_1^{\pm}(M,\kappa_1)=-iK\exp\left[\pm\frac{i\omega s}{c}+i|n|\omega\frac{\sqrt{c^2-a_1^2}}{ca_1}\pm i\chi\int_0^s\frac{ds}{\rho(s)}\right]\left\{1+\left[e_{12}(s)\left(\frac{n\omega}{a_1}\right)^2+e_{11}(s)\frac{n\omega}{a_1}+e_{10}(s)\right]\frac{a_1}{i\omega}+O(\omega^{-2})\right\}. \tag{2.3}$$

Equations (2.1), (2.2), and (2.3) apply to both Rayleigh and Stoneley waves if c is interpreted according as the solution (1.9) or the solution (1.11) of Eq. (1.5). Moreover, the following notation appears in (3.1)-(3.3):

$$c_{11}(s)=\frac{i}{\rho(s)}\left(\frac{c^2}{2(a^2-c^2)}-\frac{a\chi}{\sqrt{a^2-c^2}}\right), \tag{2.4}$$

$$d_{11}(s)=\frac{i}{\rho(s)}\left(\frac{c^2}{2(b^2-c^2)}-\frac{b\chi}{\sqrt{b^2-c^2}}\right), \tag{2.5}$$

$$e_{11}(s)=-\frac{1}{\rho(s)}\left(\frac{ic^2}{2(c^2-a_1^2)}-\frac{a_1\chi}{\sqrt{c^2-a_1^2}}\right), \tag{2.6}$$

$$\Lambda=\frac{2b^2\sqrt{a^2-c^2}}{a(2b^2-c^2)}, \tag{2.7}$$

$$K=\frac{a_1}{a}\frac{c^2}{2b^2-c^2}\frac{\sqrt{a^2-c^2}}{\sqrt{c^2-a_1^2}}, \tag{2.8}$$

$$\chi=\frac{\chi_1}{\chi_2}, \tag{2.9}$$

$$\chi_1=b(a^2-c^2)^{3/2}(2b^4-2b^2c^2+c^4)+a(b^2-c^2)^{3/2}(2a^2b^2-a^2c^2-b^2c^2+c^4)+h\left[\frac{c^4a_1\sqrt{a^2-c^2}}{2i(c^2-a_1^2)^{3/2}}-\frac{ac^4}{2(a^2-c^2)}-2b\sqrt{a^2-b^2}\sqrt{b^2-c^2}\right], \tag{2.10}$$

$$\chi_2=2b\sqrt{a^2-c^2}\sqrt{b^2-c^2}\left[b(4a^2b^2-3a^2c^2-3b^2c^2+2c^4)-2a(2b^2-c^2)\cdot\right.$$
$$\left.\cdot\sqrt{a^2-c^2}\sqrt{b^2-c^2}\right]+\frac{h}{\sqrt{a^2-c^2}}\left[\frac{c^4(a^2-a_1^2)}{c^2-a_1^2}+\frac{2b^2(6b^2+c^2)(a^2-c^2)}{2b^2-c^2}\right], \tag{2.11}$$

$$h=i\rho_0\frac{c^2a_1(a^2-c^2)(b^2-c^2)}{2b\sqrt{c^2-a_1^2}}.$$

The cumbersome explicit expressions for $c_{12}(s), d_{12}(s), e_{12}(s), c_{10}(s), d_{10}(s)$, and $e_{10}(s)$ are omitted in the interest of space.

We now examine the effect of the excitation of Rayleigh and Stoneley waves at the point M (which may be located either in the elastic medium or in the fluid medium) by a point source at $M_0(s_0,\nu_0)$ in the case of sufficiently smooth variation of the curvature of the boundary. This effect is determined by the unknown functions $\Phi_{R,S}(M_0,M,K), \Psi_{R,S}(M_0,M,\varkappa), \Phi_{1R,1S}(M_0,M,K_1)$. In order for the first equation (1.1) to be satisfied for $M\neq M_0$ the functions Φ_R and Φ_S for $s\gtrless s_0$ must be proportional to the expression $\varphi^{\pm}(M,K)$, in which $c=c_R$ or $c=c_S$, respectively. In order to guarantee the symmetry of the functions $\Phi_R(M_0,M,K)$ and $\Phi_S(M_0,M,K)$ these functions must also contain the factor $\varphi^{\mp}(M_0,K)$ for $s\gtrless s_0$. We therefore set

$$\Phi_{R,S}(M_0, M, \kappa) = A_{R,S}\,\varphi^{\mp}(M_0, \kappa)\,\varphi^{\pm}(M, \kappa)\Big|_{c=c_R, c_S}, \quad s \gtrless s_0. \tag{2.12}$$

The factors A_R and A_S are to be determined and are independent of either the coordinates of the point M or the coordinates of M_0.

The functions $\Psi_{R,S}(M_0, M, \varkappa)$ and $\Phi_{1R,1S}(M_0, M, \kappa_1)$ must, in order to satisfy the corresponding Helmholtz equations, be proportional to $\psi^{\pm}(M, \varkappa)$ or $\varphi_1^{\pm}(M, \kappa_1)$ for $s \gtrless s_0$, respectively. In order for the boundary conditions to be met, these functions, like (2.12), must also contain the factor $A_{R,S}\,\varphi^{\mp}(M_0, \kappa)$. Consequently,

$$\Psi_{R,S}(M_0, M, \varkappa) = A_{R,S}\,\varphi^{\mp}(M_0, \kappa)\,\psi^{\pm}(M, \varkappa)\Big|_{c=c_R, c_S}, \quad s \gtrless s_0, \tag{2.13}$$

$$\Phi_{1R,1S}(M_0, M, \kappa_1) = A_{R,S}\,\varphi^{\mp}(M_0, \kappa)\,\varphi_1^{\pm}(M, \kappa_1)\Big|_{c=c_R, c_S}, \quad s \gtrless s_0. \tag{2.14}$$

Equations (2.12), (2.13), and (2.14) establish the dependence of the unknown expressions $\Phi_{R,S}$, $\Psi_{R,S}$, and $\Phi_{1R,1S}$ on the coordinates of the points M and M_0

Equations (2.1), (2.2), and (2.3) make it possible to state the impedance boundary conditions that the functions $\varphi^{\pm}(M, \kappa)$, $\psi^{\pm}(M, \varkappa)$, and $\varphi_1^{\pm}(M, \kappa_1)$ must satisfy. According to (2.12), (2.13), and (2.14) these conditions also hold for the functions $\Phi_{R,S}$, $\Psi_{R,S}$, and $\Phi_{1R,1S}$. For example, the functions $\Phi_{R,S}(M_0, M, \kappa)$ satisfy the boundary condition

$$\frac{\partial \Phi_{R,S}}{\partial n} + \kappa\left(q_0 + \frac{q_1}{i\kappa} + O(\kappa^{-2})\right)\Phi_{R,S}\Big|_{n=0} = 0, \tag{2.15}$$

in which

$$q_0 = \frac{\sqrt{a^2 - c^2}}{c}\Big|_{c_R, c_S}, \qquad q_1 = -c_{11}(s)\Big|_{c_R, c_S}. \tag{2.16}$$

Consequently, the functions Φ_R and Φ_S have much in common with the function Γ_0 from [1]. These functions satisfy a boundary condition of the same genre as the function Γ_0 in § 4 of [1] and are "generated" by the inhomogeneous equation (1.1), which does not differ from the inhomogeneous "generating" equation for the function Γ_0. However, there are two essential differences: 1) The part of the Green functions $\Phi(M_0, M, \kappa)$ unrelated to surface-wave effects (unlike the analogous function Γ in [1]) does not satisfy the impedance condition (2.15); 2) two surface modes are at once generated in problem (1.1)-(1.3). These considerations prohibit the direct application of the results of [1], say, for example, the substitution of the values of q_0 and q_1 according to (2.16) into Eq. (4.3) of [1]. We can, however, determine the coefficients A_R and A_S in (2.12)-(2.14) on the basis of the same arguments as in [1].

§ 3. Determination of the Coefficients A_R and A_S

The coefficients M and M_0 in Eqs. (2.12)-(2.14), being independent of the positions of A_R and A_S, are dimensionless. This is readily confirmed by comparing Eqs. (1.1), (2.1), and (2.12). Considerations based on the localization principle (see [1], § 3) indicate that these coefficients are independent of either κ or the properties of the contour S. In view of this universal quality these coefficients can be found from the exact solution of any special problem of the type in question.

For the purpose of determining A_R and A_S we consider problem (1.1)-(1.3) (subject to the absorption limit conditions) in the case when S is the straight line $y=0$ (x and y are the Cartesian coordinates). The exact solution of the foregoing problem is elementary to formulate and takes the form

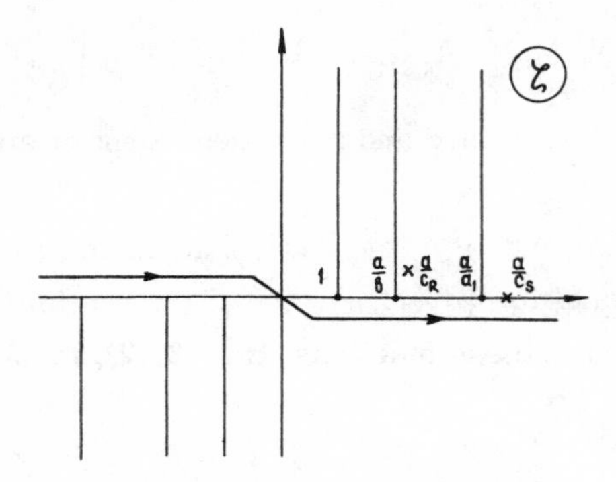

Fig. 1

$$\Phi(x_0, y_0, x, y) = \frac{1}{2\pi}\int_{-\infty}^{\infty}\left[\frac{(2\zeta_1^2 - \frac{a^2}{b^2})^2}{\sqrt{\zeta^2-1}\,\Delta(\zeta)} e^{-\eta\sqrt{\zeta^2-1}} - \frac{\operatorname{ch}(\eta\sqrt{\zeta^2-1})}{\sqrt{\zeta^2-1}}\right]\exp(-\eta_0\sqrt{\zeta^2-1} + i\kappa x\zeta)\,d\zeta, \tag{3.1}$$

$$\Psi(x_0, y_0; x, y) = \frac{1}{2\pi}\int_{-\infty}^{\infty}\frac{(-2i\zeta)(2\zeta^2 - \frac{a^2}{b^2})}{\Delta(\zeta)}\exp\left(-\eta\sqrt{\zeta^2 - \frac{a^2}{b^2}} - \eta_0\sqrt{\zeta^2-1} + i\kappa x\zeta\right)d\zeta, \tag{3.2}$$

$$\Phi_1(x_0, y_0; x, y) = \frac{1}{2\pi}\int_{-\infty}^{\infty}\frac{\frac{a^2}{b^2}(2\zeta^2 - \frac{a^2}{b^2})}{\sqrt{\zeta^2 - \frac{a^2}{b^2}}\,\Delta(\zeta)}\exp\left[-|\eta|\sqrt{\zeta^2 - \frac{a^2}{b^2}} - \eta_0\sqrt{\zeta^2-1} + i\kappa x\zeta\right]d\zeta. \tag{3.3}$$

The contour of integration in Eqs. (3.1)-(3.3) is portrayed in Fig. 1, which also shows the branch cuts $\eta = \kappa y$ and $\eta_0 = \kappa y_0$ in the ζ plane:

$$\Delta(\zeta) = \left(2\zeta^2 - \frac{a^2}{b^2}\right)^2 - 4\zeta^2\sqrt{\zeta^2-1}\sqrt{\zeta^2 - \frac{a^2}{b^2}} + \rho_0\frac{a^4}{b^4}\frac{\sqrt{\zeta^2-1}}{\sqrt{\zeta^2 - \frac{a^2}{a_1^2}}}. \tag{3.4}$$

All radicals of the form $\sqrt{\zeta^2 - \zeta_0^2}$ are assumed to be positive when $\zeta > \zeta_{10}$.

Above the contour of integration the function $\Delta(\zeta)$ in (3.1)-(3.3) has two zeros: $\zeta = \zeta_R = \frac{a}{c_R}$ and $\zeta = \zeta_S = \frac{a}{c_S}$. For $x > 0$ the integrals (3.1)-(3.3) reduce to integrals around loops enclosing the cuts plus the residues at the indicated roots. We obtain

$$\Phi = \Phi_R + \Phi_S + \tilde{\Phi}, \qquad \Psi = \Psi_R + \Psi_S + \tilde{\Psi}, \qquad \Phi_1 = \Phi_{1R} + \Phi_{1S} + \tilde{\Phi}_1. \tag{3.5}$$

Here the terms $\Phi_{R,S}$, $\Psi_{R,S}$, and $\Phi_{1R,1S}$ are induced by the residues at $\zeta = \zeta_{R,S}$, while the terms $\tilde{\Phi}$, $\tilde{\Psi}$, and $\tilde{\Phi}_1$ contain integrals around the loops enclosing the cuts in the upper half-plane of ζ. Here

$$\Phi_{R,S} = \frac{i(2\zeta^2 - \frac{a^2}{b^2})^2}{\sqrt{\zeta^2-1}\,\Delta'(\zeta)}\exp\left[-\eta_0\sqrt{\zeta^2-1} - \eta\sqrt{\zeta^2-1} + i\kappa x\zeta\right]_{\zeta_R,\zeta_S}; \tag{3.6}$$

$$\Psi_{R,S} = \frac{2\zeta(2\zeta^2 - \frac{a^2}{b^2})^2}{\Delta'(\zeta)}\exp\left[-\eta_0\sqrt{\zeta^2-1} - \eta\sqrt{\zeta^2 - \frac{a^2}{b^2}} + i\kappa x\zeta\right]_{\zeta_R,\zeta_S}; \tag{3.7}$$

$$\Phi_{1R,1S} = \frac{ia^2}{b^2}\frac{2\zeta^2 - \frac{a^2}{b^2}}{\sqrt{\zeta^2 - \frac{a^2}{b^2}}\,\Delta'(\zeta)}\exp\left[-\eta_0\sqrt{\zeta^2-1} - |\eta|\sqrt{\zeta^2 - \frac{a^2}{a_1^2}} + i\kappa x\zeta\right]_{\zeta_R,\zeta_S}. \tag{3.8}$$

The terms $\tilde{\Phi}$, $\tilde{\Psi}$, and $\tilde{\Phi}_1$ in Eqs. (3.5) characterize the point-source excitation of space waves, and Eqs. (3.6)-(3.8) determine the excitation of surface waves in the given problem. We now compare Eqs. (3.6) and (2.12), taking account of the fact that the functions $\varphi^{\pm}(M, \kappa)$ are specified by Eq. (2.1). We arrive at the following expressions for the universal coefficients:

$$A_{R,S} = \left. \frac{i\left(2\zeta^2 - \frac{a^2}{b^2}\right)^2}{\sqrt{\zeta^2 - 1}\,\Delta(\zeta)} \right|_{\zeta = \frac{a}{c_R}, \frac{a}{c_S}},$$

$$\Delta'(\zeta) = 8\zeta\left(2\zeta^2 - \frac{a^2}{b^2}\right) - 8\zeta\sqrt{\zeta^2 - 1}\sqrt{\zeta^2 - \frac{a^2}{b^2}} - 4\zeta^3 \frac{2\zeta^2 - 1 - \frac{a^2}{b^2}}{\sqrt{\zeta^2 - 1}\sqrt{\zeta^2 - \frac{a^2}{b^2}}} + \rho_0 \frac{a^4}{b^4} \zeta \frac{1 - \frac{a^2}{a_1^2}}{\sqrt{\zeta^2 - 1}\left(\zeta^2 - \frac{a^2}{a_1^2}\right)^{3/2}}. \tag{3.9}$$

Equation (3.9) also ensures the compatibility of expressions (3.7) and (3.8) with (2.13) and (2.14).

Inequalities (1.7) and (1.12) permit Eqs. (3.9) to be simplified. We have

$$A_R = \frac{1}{i} \frac{a^2 (2b^2 - c_0^2)^2}{4b^3 (2a^2 b^2 - b^2 c_0^2 - a^2 c_0^2)(b^2 - c_0^2)^{-1/2} - 2a(4b^4 - c_0^4)\sqrt{a^2 - c_0^2}} \left[1 + O(\xi)\right], \tag{3.10}$$

$$A_S = \frac{1}{i} \frac{\rho_0}{2} \frac{a^6 a_1^2}{b^2 (a^2 - b^2)^3} \left[1 + O\left(\frac{a_1^2}{b^2}\right)\right]. \tag{3.11}$$

A comparison of these equations shows that the given point source excites Rayleigh waves far more intensely than Stoneley waves:

$$\frac{A_S}{A_R} = O\left(\frac{\rho_0^2 a_1^2}{b^2}\right).$$

§ 4. Final Equations and Their Analysis

The high-frequency asymptotic equations for $\Phi_{R,S}$, $\Psi_{R,S}$, and $\Phi_{1R,1S}$ are obtained by the substitution of expressions (2.1)-(2.3) and (3.9) into (2.12)-(2.14). These equations give the solution of the stated problem of the excitation of Rayleigh and Stoneley waves by a point source. We shall concern ourselves with the most interesting case, in which the points M_0 and M are situated close to the boundary S (for finite values of ν, $\bar{\nu}$, and $|\nu_1|$). In order to imbue these equations with a more transparent form we simplify them on the basis of inequality (1.7) in the Rayleigh-wave case and on the basis of inequality (1.12) in the Stoneley-wave case.

Damped Rayleigh Waves. We obtain the following:

$$\Phi_R(M_0, M, \kappa) = A_R \exp\left[\frac{i\omega|s - s_0|}{c_R} - (\nu + \nu_0)\frac{\sqrt{a^2 - c_0^2}}{c_0} + i\chi_R \int_{s_<}^{s_>} \frac{ds}{\rho(s)}\right]\left[1 + O(\omega^{-1}) + O(\xi)\right], \tag{4.1}$$

$$\Psi_R(M_0, M, \varkappa) = \mp i\Lambda A_R \exp\left[\frac{i\omega|s - s_0|}{c_R} - \frac{\nu_0\sqrt{a^2 - c_0^2} + \bar{\nu}\sqrt{b^2 - c_0^2}}{c_0} + i\chi_R \int_{s_<}^{s_>} \frac{ds}{\rho(s)}\right]\left[1 + O(\omega^{-1}) + O(\xi)\right], \quad s \gtrless s_0, \tag{4.2}$$

$$\Phi_{1R}(M_0, M, \kappa_1) = -iKA_R \exp\left[\frac{i\omega|s - s_0|}{c_R} - \nu_0\frac{\sqrt{a^2 - c_0^2}}{c_0} + i|\nu_1|\left(1 - i\varepsilon - \frac{1}{2}\frac{a_1^2}{c_0^2}\right) + i\chi_R \int_{s_<}^{s_>} \frac{ds}{\rho(s)}\right]\left[1 + O(\omega^{-1}) + O(\xi)\right], \tag{4.3}$$

$$M_0 = (s_0, n_0), \quad M = (s, n), \quad s_> = \max(s, s_0), \quad s_< = \min(s, s_0).$$

The expressions A_R, Λ, K, and ξ involved here are determined by Eqs. (3.10), (2.7), (2.8), and (1.7). The equation for χ_R is obtained from Eqs. (2.9)-(2.11) for χ with the substitution therein of $h=0$.

Using (4.1)-(4.3), we can write an expression for the Rayleigh-wave phase velocity v_R. We obtain for $\omega\to\infty$

$$v_R(s)=\frac{1}{\mathrm{Re}(c_R^{-1})}-\frac{\chi_R}{[\mathrm{Re}(c_R^{-1})]^2\omega\rho(s)}+O(\omega^{-2})=c_0-\frac{c_0^2\chi_R}{\omega\rho(s)}+O(\xi^2)+O(\xi\omega^{-1})+O(\omega^{-2}). \tag{4.4}$$

It can be shown on the basis of Eq. (1.8) that $\chi_R>0$. If $\rho^{-1}(s)>0$ (contour S concave relative to the elastic medium), the dispersion is anomalous. If $\rho^{-1}(s)<0$, the dispersion is normal.

Other characteristics of the Rayleigh wave are the damping scales l_φ and l_ψ of the potentials Φ_R and Ψ_R with distance from the boundary S (the distances over which the amplitudes are diminished by e). We have from (4.1) and (4.2)

$$l_\varphi=\frac{ac_0}{\omega\sqrt{a^2-c_0^2}},\qquad l_\psi=\frac{bc_0}{\omega\sqrt{b^2-c_0^2}}>l_\varphi. \tag{4.5}$$

The limiting values of the field Φ_{1R} on S is smaller by $O\left(\frac{a_1}{b}\right)$ than the analogous values of the fields Φ_R and Ψ_R. With distance from S, however, the field Φ_{1R} does not decay, but oscillates with a slowly increasing amplitude, in correspondence with the factor $\exp(i|\nu_1|+\varepsilon|\nu_1|)$. Finally, all three functions Φ_R, Ψ_R, and Φ_{1R} decay as $|s-s_0|$ increases. The scale L of this decay is of the order

$$L=O\left(\frac{b^2}{\omega\rho_0 a_1}\right). \tag{4.6}$$

Undamped Rayleigh Waves. Passage to the limit as $\rho_0\to 0$ and $\frac{a_1}{b}\to 0$ in Eqs. (4.1)-(4.3) corresponds to the excitation of Rayleigh waves at an interface between an elastic medium and a vacuum. The potential Φ_1 vanishes in this case, the third boundary condition (1.3) is automatically satisfied, and Eq. (1.5) for the velocity c reverts to the Rayleigh equation (1.8). The Rayleigh waves turn out to be undamped along the coordinate s. Their generation is described by Eqs. (4.1) and (4.2), in which it is required to set $c_R=c_0$ and $\xi=0$. Equation (4.4) experiences an analogous change, while (4.5) remains intact.

For the case of an interface between an elastic medium and a vacuum the functions $\varphi^{\pm}(M,\kappa)$ and $\psi^{\pm}(M,\varkappa)$ have been determined in [4] in the very general case of the three-dimensional problem and an arbitrary smooth dependence of the velocities a and b on the coordinates. Using these solutions and finding the "spatial" coefficient A_R, we can use Eqs. (2.12) and (2.13) to describe the excitation of Rayleigh waves at a boundary with a vacuum in the indicated general case.

Stoneley Waves. We now substitute (2.1)-(2.3) into (2.12)-(2.14) with regard for inequalities (1.12). In Eqs. (2.10) and (2.11), which determine χ, the principal terms are those containing the difference $c^2-a_1^2$ in the denominator. After some obvious simplifications we obtain

$$\Phi_s(M_0,M,\kappa)=A_s\exp\left[\frac{i\omega|s-s_0|}{a_1(1-\delta)}-\frac{a(\nu_0+\nu)}{a_1}+i\chi_s\int_{s_<}^{s_>}\frac{ds}{\rho(s)}\right]\left[1+O(\omega^{-1})+O\left(\frac{a_1^2}{b^2}\right)\right], \tag{4.7}$$

$$\Psi_s(M_0,M,\varkappa)=\mp iA_s\exp\left[\frac{i\omega|s-s_0|}{a_1(1-\delta)}-\frac{a\nu_0+b\bar{\nu}}{a_1}+i\chi_s\int_{s_<}^{s_>}\frac{ds}{\rho(s)}\right]\left[1+O(\omega^{-1})+O\left(\frac{a_1^2}{b^2}\right)\right],\quad s\gtrless s_0, \tag{4.8}$$

$$\Phi_{1s}(M_0, M, \kappa_1) = -\frac{(a^2-b^2)}{\rho_0 a^2} A_s \exp\left[\frac{i\omega|s-s_0|}{a_1(1-\delta)} - \frac{a\nu_0}{a_1} - |\nu_1|\sqrt{2\delta} + i\chi_s \int_{s_<}^{s_>} \frac{ds}{\rho(s)}\right]\left[1 + O(\omega^{-1}) + O\left(\frac{a_1^2}{b^2}\right)\right], \tag{4.9}$$

$$\chi_s = \frac{1}{2\sqrt{2\delta}}\left[1 + O(\delta)\right] = \frac{b^2(a^2-b^2)}{\rho_0 a^2 a_1^2}\left[1 + O\left(\frac{a_1^2}{b^2}\right)\right],$$

in which A_s and δ are determined by Eqs. (3.11) and (1.13).

We deduce the following results on the basis of Eqs. (4.7)-(4.9). The Stoneley-wave phase velocity v_s is equal to

$$v_s(s) = a_1(1-\delta) - \frac{a_1^2 \chi_s}{\omega \rho(s)} + O(\omega^{-2}) + O(\delta\omega^{-1}). \tag{4.10}$$

As in the Rayleigh-wave case, anomalous dispersion prevails for $\rho^{-1}(s) > 0$, while for $\rho^{-1}(s) < 0$ normal dispersion takes place. Inasmuch as $\chi_s \gg 1$, the dispersion in the case of Stoneley waves is considerably more pronounced than in the case of Rayleigh waves.

At the boundary S the amplitude of the field Φ_{1s} is greater by $O(\rho_0^{-1})$ than the amplitudes of the fields Φ_s and Ψ_s. The Stoneley wave propagates along S without damping. The damping scales l_φ, l_ψ, and $l_{1\varphi}$ of the potentials Φ_s, Ψ_s, and Φ_{1s} with distance from S are as follows:

$$l_\varphi = l_\psi = \frac{a_1}{\omega}, \qquad l_{1\varphi} = \frac{a_1}{\omega\sqrt{2\delta}}, \tag{4.11}$$

$$l_{1\varphi} \gg l_\varphi.$$

Thus, the Stoneley wave is concentrated predominantly in the liquid (gas).

LITERATURE CITED

1. Molotkov, I. A., Surface-wave excitation in connection with diffraction at an impedance contour, this volume, pp. 83-92.
2. Brekhovskikh, L. M., Waves in Layered Media, Izd. AN SSSR (1956) [English translation: Academic Press, New York (1960)].
3. Stoneley, R., The effect of the ocean on Rayleigh waves, Month. Not. Roy. Astronom. Soc., Geophys. Suppl., 1:349 (1926).
4. Babich, V. M. and Rusakova, N. Ya., Propagation of Rayleigh waves over the surface of an inhomogeneous elastic body of arbitrary configuration, Zh. Vychis. Matem. i Matem. Fiz., 2(4): 652-665 (1962).

ON THE EIGENMODES OF A THREE-DIMENSIONAL MULTILAYER RESONATOR

T. F. Pankratova

In the present article we investigate the stability of a system of rays in the vicinity of the axis of a three-dimensional multilayered resonator, formulate a solution of the Maxwell equations by the parabolic equation method, and derive an approximate equation for the eigenfrequencies. It turns out that the multilayered resonator is stable only if "compatibility conditions" are imposed on the resonator parameters. The case in which the problem is reducible to a two-dimensional problem and the wave propagation velocity in each layer is constant has been investigated in [1]. In that article, however, essential use was made of the two-dimensionality of the problem, so that the analogous investigation of a three-dimensional resonator has necessitated a special study. The present investigation relies heavily on the notions of [1], abetted by consultations and suggestions from its author, Prof. V. M. Babich, to whom I express my sincere appreciation.

§1. Ray Formulation of the Problem

Let us consider the problem of the oscillatory modes in a resonator synthesized from two perfectly reflecting mirrors and filled with a layered medium. We regard each layer as inhomogeneous and isotropic; the wave velocity changes discontinuously at the boundary. We define the rays as extremals of the functional $\gamma = \int_{M_0}^{M} \frac{d\sigma}{v(\vec{r})}$, where r is the radius vector of a variable point and $d\sigma$ is an element of length. We assume that there is a certain extremal (the resonator axis), near which the electromagnetic oscillations are concentrated. As is so often done (see [2, 3]), we introduce in the vicinity of the axis in each jth layer an orthogonal coordinate system (s_j, ξ_j, η_j), where s_j is measured along the extremal and ξ_j, and η_j lie in a plane perpendicular to it and are rotated through an angle $\theta_j = \int_{\theta_j^0}^{\theta} \varkappa_j \, ds$ with respect to the normal and binormal, where $\varkappa_j$ is the torsion of the axis. The specification of each of these coordinate systems has an arbitrariness in the choice of reference origin for the arclength s_j^0 and initial angle θ_j^0. We choose these arbitrary constants so that one coordinate system merges continuously into the next in going from the jth medium to the $(j+1)$th medium, i.e., so that

$$s_j \big|_{S_j} = s_{j+1}^0 \big|_{S_j} \equiv \ell_j , \tag{1.1}$$

$$\theta_j \big|_{S_j} = \theta_{j+1}^0 \big|_{S_j} , \tag{1.2}$$

where S_j is the interface between the jth and $(j+1)$th media.

The surfaces of the mirrors terminating the $\mathcal{N}$-layer resonator are denoted by S_0 and $S_{\mathcal{N}}$. At each point of intersection of the resonator axis with the surface S_j we can place a Cartesian coordinate system $(\zeta_{1j}, \zeta_{2j}, \zeta_{3j})$ so that the axis ζ_{3j} is perpendicular to the surface, while ζ_{1j} and ζ_{2j} lie in the tangent plane. We shall assume that the equation for the surface S_j has the form

$$\zeta_{3j} = (D^{(j)} \vec{\zeta}_j, \vec{\zeta}_j), \quad j = 0, 1, \ldots, N, \tag{1.3}$$

in which

$$\vec{\zeta}_j = (\zeta_{1j}, \zeta_{2j}), \qquad D^{(j)} = \begin{pmatrix} D_{11}^{(j)} & D_{12}^{(j)} \\ D_{12}^{(j)} & D_{22}^{(j)} \end{pmatrix}.$$

We describe the rays in the vicinity of the resonator axis approximately by means of the Euler equations for the functional J_0, which is equal to the functional J in the vicinity of the axis up to squared terms (see [3]):

$$J = J_0 + O(\xi^3, \eta^3, \ldots). \tag{1.4}$$

The Euler equations for J_0 have the following form in the coordinates (s_j, ξ_j, η_j) (see [3]):

$$\frac{d^2\vec{\gamma}_j}{ds^2} - \frac{v'(s)}{v(s)} \frac{d\vec{\gamma}_j}{ds} + K(s)\vec{\gamma}_j = 0, \tag{1.5}$$

where

$$\vec{\gamma}_j(s) = (\gamma_{j\xi}(s), \gamma_{j\eta}(s)), \quad v'(s) = \left.\frac{dv}{ds}\right|_{s,0,0},$$

$$K(s) = \frac{1}{v(s)} \begin{pmatrix} \left.\frac{\partial^2 v}{\partial \xi^2}\right|_{s,0,0} & \left.\frac{\partial^2 v}{\partial \xi \partial \eta}\right|_{s,0,0} \\ \left.\frac{\partial^2 v}{\partial \xi \partial \eta}\right|_{s,0,0} & \left.\frac{\partial^2 v}{\partial \eta^2}\right|_{s,0,0} \end{pmatrix}.$$

The vector $\vec{\gamma}_j(s_j)$, satisfying the system (1.5) characterizes a certain ray in the jth layer of the resonator. In each layer we fix four linearly independent solutions (forming a basis) of Eqs. (1.5): $\vec{\gamma}_1^{(j)}, \vec{\gamma}_2^{(j)}, \vec{\gamma}_3^{(j)}, \vec{\gamma}_4^{(j)}$ Now the ray can be characterized by a constant vector independent of the length s_j. Thus, any solution of the system (1.5) is representable in the form

$$\vec{\gamma}_j(s_j) = \sum_{K=1}^{4} a_K^{(j)} \vec{\gamma}_K^{(j)}, \tag{1.6}$$

i.e., the variable vector $\vec{\gamma}_j(s_j)$ is determined by the constant vector $\vec{a}_j = (a_1^{(j)}, a_2^{(j)}, a_3^{(j)}, a_4^{(j)})$ with given $\vec{\gamma}_K^{(j)}$, $K=1, 2, 3, 4$.

Let the ray described by the vector $\vec{a}_{j1}$ impinge at a small angle on the surface S_j (in the vicinity of the point of intersection of the axis with the surface):

$$\vec{\gamma}_{j1}(s_j) = \sum_{K=1}^{4} a_{K1}^{(j)} \vec{\gamma}_K^{(j)}(s_j).$$

The reflected (refracted) ray is characterized by the vector $\vec{a}_{j2}(\vec{a}_{(j+1)1})$:

$$\vec{\gamma}_{j2}(s_j)=\sum_{K=1}^{4} a^{(j)}_{K2}\vec{\gamma}^{(j)}_{K}(s_j) \qquad (\vec{\gamma}_{(j+1)1}(s_j)=\sum_{K=1}^{4} a^{(j+1)}_{K1}\vec{\gamma}^{(j+1)}_{K}(s_j))$$

The vector $\vec{a}_{j2}(\vec{a}_{(j+1)1})$ is related to the vector $\vec{a}_{j1}$ by the reflection matrix Γ_j (refraction matrix B_j):

$$\vec{a}_{j2}=\Gamma_j\vec{a}_{j1} \quad (\vec{a}_{(j+1)1}=B_j\vec{a}_{j1}), \tag{1.7}$$

in which

$$\Gamma_j = w_j^{-1}(\ell_j)u^{-1}(\theta_j)g_j u(\theta_j)w_j(\ell_j),$$

$$B_j = w_{j+1}^{-1}(\ell_j)u^{-1}(\theta_j)b_j(u(\theta_j)w_{j+1}(\ell_j),$$

$$g_j=\begin{pmatrix} 1 & 0 & 0 & 0 \\ -4D_{11}^{(j)} & -1 & -4D_{12}^{(j)} & 0 \\ 0 & 0 & 1 & 0 \\ -4D_{12}^{(j)} & 0 & -4D_{22}^{(j)} & -1 \end{pmatrix}, \quad b_j=\begin{pmatrix} 1 & 0 & 0 & 0 \\ -2\left(1-\frac{v_{j+1}}{v_j}\right)D_{11}^{(j)} & \frac{v_{j+1}}{v_j} & -2\left(1-\frac{v_{j+1}}{v_j}\right)D_{12}^{(j)} & 0 \\ 0 & 0 & 1 & 0 \\ -2\left(1-\frac{v_{j+1}}{v_j}\right)D_{12}^{(j)} & 0 & -2\left(1-\frac{v_{j+1}}{v_j}\right)D_{22}^{(j)} & \frac{v_{j+1}}{v_j} \end{pmatrix}.$$

$D_{iK}^{(j)}$ are the coefficients of the quadratic form (1.3), $w_j(s)$ is a matrix, each column of which has the form

$$\begin{pmatrix} \gamma^{(j)}_{K\xi} \\ \frac{d}{ds}\gamma^{(j)}_{K\xi} \\ \gamma^{(j)}_{K\eta} \\ \frac{d}{ds}\gamma^{(j)}_{K\eta} \end{pmatrix}, \qquad K=1,2,3,4,$$

and the matrix $u(\theta_j)$ is determined by the rotation of the coordinates ξ_1 and ξ_2 with respect to ξ_j and η_j at the point of incidence. It may be assumed without loss of generality that $u(\theta_j)=I$ (the unit matrix). The reflection matrix Γ_j is found in the form (1.7) in [2]. The expressions for B_j are deduced analogously by the techniques of the variational calculus.

We choose the basis in each layer so that $\vec{\gamma}_K^{(j+1)}(\ell_j)=\vec{\gamma}_K^{j}(\ell_j)$ $\frac{d}{ds}\vec{\gamma}_K^{j+1}(\ell_j)=\frac{d}{ds}\vec{\gamma}_K^{j}(\ell_j)$. Then it is easily shown that the matrices Γ_j and B_j have the following properties:

$$\Gamma_j=\Gamma_j^{-1}, \tag{1.8}$$

$$\Gamma_j B_j = B_j\Gamma_j, \tag{1.9}$$

$$\det\Gamma_j=-1. \tag{1.10}$$

In the first layer we fix a certain ray, determined by the vector $\vec{a}_1$. After several reflections and refractions it is characterized by the vector

$$\vec{a}_\Phi=\Phi\vec{a}_1,$$

where the matrix Φ is equal to a product of matrices Γ_j, B_j, and B_j^{-1}; for the ray illustrated in Fig. 1, for example,

$$\Phi=B_1^{-1}\Gamma_2\Gamma_1\Gamma_2B_1\Gamma_0\Gamma_1\Gamma_0.$$

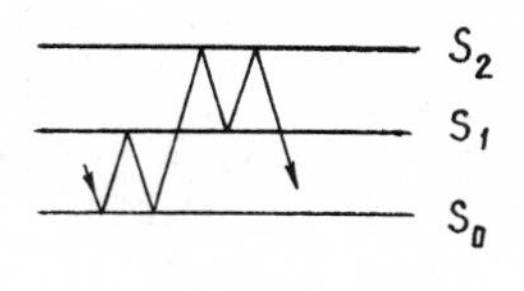

Fig. 1

Each ray can be determined in terms of the vector $\vec{a}_1$ by means of its particular matrix Φ. Following [1], we say that the resonator is stable in the first approximation if all the matrices Φ corresponding to rays entering the resonator have norms* bounded in the aggregate, i.e., if there is a number K such that for any matrix Φ

$$\|\Phi\| \leqslant K .$$

This definition may be confined to rays entering the first layer "from below" (see [1]). We now derive necessary and sufficient conditions for the stability of a multilayered resonator.

§ 2. Necessary and Sufficient Conditions for Stability

It is readily seen, by analogy with the two-dimensional case (see [1]), that the set M of matrices Φ ($\{\Phi\} \equiv M$) forms a monoid, i.e., a set in which:

1) multiplication is defined and

$$\Phi_1(\Phi_2\Phi_3) = (\Phi_1\Phi_2)\Phi_3 \quad \text{for any} \quad \Phi_1, \Phi_2, \Phi_3, \tag{2.1}$$

2) there is an identity element: $e\Phi = \Phi e = \Phi$.

The determinant of any matrix $\Phi \in M$ is equal to unity:

$$\det \Phi = 1, \tag{2.2}$$

i.e., the matrices Φ are unimodular.

The problem of the resonator stability conditions reduces to the following: What must the monoid M, consisting of 4×4 unimodular matrices Φ be in order that for all $\Phi \in M$

$$\|\Phi\| \leqslant K \quad (K \text{ independent of } \Phi)? \tag{2.3}$$

The following is true.

Theorem 1. *In order for the unimodular matrices $\Phi \in M$ to have the property (2.3) it is necessary and sufficient that there exist a real matrix C, $\det C \neq 0$, independent of Φ, such that*

$$\Phi = C U_\Phi C^{-1}, \tag{2.4}$$

where U_Φ is an orthogonal matrix (for each member Φ).

The sufficiency clause for matrices of any order is obvious:

$$\|\Phi\| = \|C U_\Phi C^{-1}\| \leqslant \|C\| \, \|U_\Phi\| \, \|C^{-1}\| = \|C\| \, \|C^{-1}\| .$$

The necessity clause is easily deduced by complementing the set M to a group. To do this we first prove the following lemma.

Lemma. *Let A be a square unimodular real quasi-orthogonal fourth-order matrix:*

*The norm of the matrices Φ is interpreted in the customary sense as the minimum number of sets of numbers c such that $|\Phi\vec{X}| \leqslant c\,|\vec{X}|$, where $|\vec{f}|$ is the length of the vector $\vec{f}$.

$$A = C_u \, U \, C_u^{-1} \quad (\det C_u \neq 0);$$

then there is a sequence of natural numbers $\{\tau_j\}$ such that in the norm

$$A^{\tau_j} \underset{j\to\infty}{\longrightarrow} A^{-1}. \tag{2.5}$$

Proof. Let $|\theta|$ be the distance from the number θ to the nearest integer. The following theorem holds (see [4]):

There is an infinite set $\tilde{R}$ of integral solutions of the inequality

$$\tilde{\tau}^{\frac{1}{n}}(\max|\tilde{\tau}\theta_1|, |\tilde{\tau}\theta_2|, \ldots, |\tilde{\tau}\theta_n|) < \frac{n}{n+1}, \tag{2.6}$$

where $\theta_1, \theta_2, \ldots, \theta_n$ are any real numbers.

We need only consider $\tilde{\tau} > 0$ since $\tilde{\tau} = 0$ yields a trivial result, whereas $\tilde{\tau} < 0$ does not generalize the situation ($|\tilde{\tau}\theta| = |(-\tilde{\tau})\theta|$). From the solutions of inequality (2.6) we can form a sequence $\{\tilde{\tau}_j\}$ of natural numbers for which (for $n = 2$)

$$|\tilde{\tau}_j \theta_i| \underset{j\to\infty}{\longrightarrow} 0, \qquad i = 1, 2. \tag{2.7}$$

We now show that this theorem implies the statement of the lemma (for matrices of an order n). Each orthogonal matrix U is reduced to diagonal form by a unitary transformation $\mathcal{D}$ (see [5]):

$$U = \mathcal{D}[e^{i\alpha_1}, e^{-i\alpha_1}, e^{i\alpha_2}, e^{-i\alpha_2}]\,\mathcal{D}^{-1}, \tag{2.8}$$

in which α_1 and α_2 are real numbers and $[\lambda_1, \lambda_2, \lambda_3, \lambda_4]$ is a diagonal matrix. It follows from the above that the matrix A is representable in the form

$$A = \tilde{C}\,[e^{i\alpha_1}, e^{-i\alpha_1}, e^{i\alpha_2}, e^{-i\alpha_2}]\,\tilde{C}^{-1}, \tag{2.9}$$

where $\tilde{C} = C_u \mathcal{D}$, $\det \tilde{C} \neq 0$, and

$$A^{\tau_j} = \tilde{C}\,[e^{i\tau_j\alpha_1}, e^{-i\tau_j\alpha_1}, e^{i\tau_j\alpha_2}, e^{-i\tau_j\alpha_2}]\,\tilde{C}^{-1}. \tag{2.10}$$

We denote $\frac{\alpha_i}{2\pi} \equiv \theta_i$, $i = 1, 2$.

It follows from Eq. (2.10) that if we find a sequence of natural numbers $\{\tau_j\}$ for which

$$|\tau_j\theta_i - (1 - \theta_i)| \underset{j\to\infty}{\longrightarrow} 0, \qquad i = 1, 2, \tag{2.11}$$

then for this same sequence

$$A^{\tau_j} \underset{j\to\infty}{\longrightarrow} A^{-1}. \tag{2.12}$$

But such a sequence does exist, viz., $\tau_j = \tilde{\tau}_j - 1$, where $\{\tilde{\tau}_j\}$ is the sequence (2.7).

Thus,

$$|\tau_j\theta_i - (1-\theta_i)| = |(\tau_j+1)\theta_i - 1| = |(\tau_j+1)\theta_i| = |\tilde{\tau}_j\theta_i| \underset{j\to\infty}{\longrightarrow} 0. \tag{2.13}$$

In the special case in which θ_1 and θ_2 are rational numbers, say $\theta_i = \frac{p_i}{q_i}$, $i = 1, 2$, is an irreducible fraction, the following is true for the sequence $\tau_j = j\,q_1 q_2 - 1$:

$$|(\tau_j+1)\,\theta_i\,| = 0, \qquad j = 1, 2, 3, \ldots;$$

$$A^{\tau_j} = A^{-1}, \qquad j = 1, 2, 3, \ldots.$$

Now, once the lemma has been proved, the necessity of condition (2.4) is proved by the method of [1].

Let the matrices Φ of the monoid M be bounded in the norm by the same number K in each case. Then any matrix Φ is similar to a certain orthogonal matrix

$$\Phi = C_\Phi U_\Phi C_\Phi^{-1} \tag{2.14}$$

[in fact, condition (2.3) implies that the eigenvalues of any matrix Φ are equal to unity in the modulus and the elementary divisors are prime (see [6, 7]; (2.14) is implied by this result and the fact that all Φ are real].

The set of products of the form

$$\Psi = \Phi_1^{q_1} \Phi_2^{q_2} \cdots \Phi_m^{q_m}, \quad \Phi_j \in M \tag{2.15}$$

generated by the matrices Φ_j of the monoid M form a group $\mathcal{N}$. All matrices of $\mathcal{N}$ satisfy inequality (2.3) (see [1]). The closure $\bar{\mathcal{N}}$ of $\mathcal{N}$ (limits in the norm of all subsequences $\Psi \in \mathcal{N}$) is a compact group of real matrices. It follows from the theorem of complete reducibility of the representations of compact groups (see [8]) that there is a real matrix C, identical for all $\Psi \in \bar{\mathcal{N}}$, such that

$$\Psi = C U_\Psi C^{-1}.$$

Inasmuch as $M \subset \bar{\mathcal{N}}$, the proof of Theorem 1 is complete.

Consider a two-layer resonator. We introduce the notation

$$B_\ell^{-1} \Gamma_k B_\ell \equiv \Gamma_k^{B_\ell}. \tag{2.16}$$

Any matrix Φ characterizing rays entering the first medium from below can be represented in the matrix product form

$$\Gamma_1\Gamma_0, \quad \Gamma_2^{B_1}\Gamma_0, \quad \text{and} \quad \Gamma_2^{B_1}\Gamma_1^{B_1}. \tag{2.17}$$

For stability of the resonator, therefore, it is necessary and sufficient that there exist a matrix C, $(\det C \neq 0)$ for which

$$\Gamma_1\Gamma_0 = C U_{10} C^{-1}, \tag{2.18}$$

$$\Gamma_2^{B_1}\Gamma_0 = C U_{20} C^{-1}, \tag{2.19}$$

$$\Gamma_2^{B_1}\Gamma_1^{B_1} = C U_{21} C^{-1}, \tag{2.20}$$

where U_{ik} are different orthogonal matrices.

Actually only two of the conditions (2.18)-(2.20) are necessary and sufficient, because it is readily inferred from Eqs. (1.8)-(1.10) that

$$\Gamma_1^{B_1}\Gamma_1 = I \tag{2.21}$$

and

$$\left(\Gamma_2^{B_1}\Gamma_1^{B_1}\right)^{-1}\left(\Gamma_2^{B_1}\Gamma_0\right)\left(\Gamma_1\Gamma_0\right)^{-1} = \Gamma_1^{B_1}\Gamma_1 = I,$$

where I is the unit matrix, so that the fulfillment of any two relations from (2.18)-(2.20) implies fulfillment of the third.

Analogously, for the stability of a three-layer resonator it is necessary and sufficient to augment any two conditions of (2.18)-(2.20) with a third, for example,

$$\Gamma_3^{B_2}\Gamma_2^{B_2} = CU_{32}C^{-1}. \tag{2.22}$$

Theorem 1 implies that the rays in stable resonators form closed congruences in the first medium by the vectors $\vec{a}_1$, for which

$$(c^{-1}\vec{a}_1, c^{-1}\vec{a}_1) = \varkappa, \tag{2.23}$$

where $\varkappa > 0$ is an arbitrary positive number and c is a matrix reducing all the Φ to orthogonal matrices.

After a certain number of reflections and refractions the rays from this set return to the first medium and are characterized by vectors $\vec{a}_\Phi = \Phi\vec{a}_1$, where

$$(c^{-1}\vec{a}_\Phi, c^{-1}\vec{a}_\Phi) = (c^{-1}\Phi\vec{a}_1, c^{-1}\Phi\vec{a}_1) = (u_\Phi c^{-1}\vec{a}_1, u_\Phi c^{-1}\vec{a}_1) = \varkappa,$$

i.e., they belong to the same set (2.23).

After refraction of the rays (2.23) into the second and then into the third medium the vectors

$$\vec{a}_2 = B_1\vec{a}_1, \tag{2.24}$$

$$\vec{a}_3 = B_2 B_1 \vec{a}_1 \tag{2.25}$$

are bound by the relations

$$(c^{-1}B_1^{-1}\vec{a}_2, c^{-1}B_1^{-1}\vec{a}_2) = \varkappa, \tag{2.26}$$

$$(c^{-1}B_1^{-1}B_2^{-1}\vec{a}_3, c^{-1}B_1^{-1}B_2^{-1}\vec{a}_3) = \varkappa. \tag{2.27}$$

The congruences of the rays (2.26) and (2.27) are, like (2.23), also invariant.

§ 3. Determination of the Field in the Vicinity of the Axis by the Parabolic Equation Method; Frequency Equation

Let the field in each medium be described by the system of Maxwell equations

$$\begin{cases} \operatorname{rot}\vec{E}_j = i\omega\mu_j\vec{H}_j, \\ \operatorname{rot}\vec{H}_j = -i\omega\varepsilon_j\vec{E}_j, \end{cases} \tag{3.1}$$

where $\varepsilon_j(\vec{r})$ and $\mu_j(\vec{r})$ are the dielectric constant and permeability of the jth medium. The conditions of continuity of the tangential components of $\vec{E}$ and $\vec{H}$ hold at the interface:

$$\begin{aligned} \vec{E}_{t_j}\big|_{S_j} &= \vec{E}_{t_{j+1}}\big|_{S_j}, \\ \vec{H}_{t_j}\big|_{S_j} &= \vec{H}_{t_{j+1}}\big|_{S_j}, \end{aligned} \tag{3.2}$$

and the conditions of ideal reflection hold at the mirror surfaces:

$$\vec{E}_{t_1}|_{S_0} = 0,$$
$$\vec{H}_{t_1}|_{S_0} = 0,$$
$$\vec{E}_{t_N}|_{S_N} = 0, \qquad (3.3)$$
$$\vec{H}_{t_N}|_{S_N} = 0.$$

Treating the frequency ω in Eqs. (3.1) as a large parameter, we seek solutions of (3.1) concentrated in the vicinity of the resonator axis, i.e.,

$$\max\{|\vec{E}|, |\vec{H}|\} \xrightarrow[\sqrt{\omega} R \to \infty]{} 0, \qquad (3.4)$$

where R is the distance from the axis.

We write the Maxwell equations in the coordinates (s_j, ξ_j, η_j) and seek the fields $\vec{E}$ and $\vec{H}$ in the jth medium as the superposition of two waves: the wave $\vec{E}^+$ (or $\vec{H}^+$) propagating "upward," and the wave $\vec{E}^-$ (or $\vec{H}^-$) propagating "downward":

$$\vec{E}_j^{\pm} = \frac{1}{\sqrt{\varepsilon}} \vec{\mathcal{E}}(s, \xi, \eta) \exp\left[\pm i\omega \int_{s_0}^{s} \frac{ds}{v(s, 0, 0)}\right], \qquad (3.5)$$

$$\vec{H}_j^{\pm} = \frac{1}{\sqrt{\mu}} \vec{\mathcal{H}}(s, \xi, \eta) \exp\left[\pm i\omega \int_{s_0}^{s} \frac{ds}{v(s, 0, 0)}\right], \qquad (3.6)$$

$$\vec{E} = (E_s, E_\xi, E_\eta), \quad \vec{H} = (H_s, H_\xi, H_\eta).$$

Substituting expressions (3.5) and (3.6) into Eqs. (3.1) and assuming $f = O(1)$, $\frac{\partial^{p+q}}{\partial \xi^p \partial \eta^q} f = O(\omega^{\frac{p+q}{2}})$, where f is either of the components $\vec{E}$ or $\vec{H}$, we obtain, up to terms of order $\omega^{-\frac{1}{2}}$ (see [3])

$$E_s^{\pm} = H_s^{\pm} = 0, \qquad (3.7)$$

$$\begin{cases} \mathcal{E}_\xi^+ = \mathcal{H}_\eta^+, \\ \mathcal{E}_\eta^+ = -\mathcal{H}_\xi, \end{cases} \qquad (3.8)$$

$$\begin{cases} \mathcal{E}_\xi^- = -\mathcal{H}_\eta^-, \\ \mathcal{E}_\eta^- = \mathcal{H}_\xi^-, \end{cases} \qquad (3.9)$$

$$\mathcal{E}_\kappa^{\pm} = \sqrt{v(s, 0, 0)}\, \varphi_\kappa^{\pm}, \quad \kappa = \xi, \eta, \qquad (3.10)$$

where the functions $\varphi_\xi^\pm$ and $\varphi_\eta^\pm$ are solutions of the independent parabolic equations

$$L^\pm \varphi_\kappa^\pm = 0,$$

$$L^\pm \equiv \frac{\partial^2}{\partial x^2} + \frac{\partial^2}{\partial y^2} \pm 2\frac{i}{v(s,0,0)}\frac{\partial}{\partial s} - \frac{1}{v^2(s,0,0)}(\mathcal{K}(s)\vec{x},\vec{x}), \tag{3.11}$$

in which

$$x = \sqrt{\omega}\,\xi, \quad y = \sqrt{\omega}\,\eta, \quad \vec{x} = (x, y),$$

and $\mathcal{K}(s)$ is the matrix determined by Eq. (1.5).

Let us suppose that each resonator formed in the jth medium between two adjacent boundaries S_{j-1} and S_j is stable and that the solution of the parabolic equation can be formulated by the method given in [6] (the system of Maxwell equations has been solved by this method in [3]).

Let the ray $\vec{\gamma}_{jk}^{\,-}$ impinge on the boundary S_{j-1}, and let the ray $\vec{\gamma}_{jk}^{\,+}$ impinge on S_j and, on reflection from S_j, be converted into $\vec{\gamma}_{jk}^{\,=}$, $k = 1, 2$. These three rays are characterized by the vectors $\vec{a}_{jk}^{\,-}$, $\vec{a}_{jk}^{\,+}$, and $\vec{a}_{jk}^{\,=}$ (see §1), where

$$\vec{a}_{jk}^{\,+} = \Gamma_{j-1}\vec{a}_{jk}^{\,-}, \quad \vec{a}_{jk}^{\,=} = \Gamma_j\Gamma_{j-1}\vec{a}_{jk}^{\,-}. \tag{3.12}$$

For the vector $\vec{a}_{jk}^{\,-}$ we pick the eigenvector of the matrix $\Gamma_j\Gamma_{j-1}$

$$\vec{a}_{jk}^{\,=} = \Gamma_j\Gamma_{j-1}\vec{a}_{jk}^{\,-} = \lambda\vec{a}_{jk}^{\,-}, \quad \lambda_k = e^{\pm i\alpha_k}, \quad k = 1, 2. \tag{3.13}$$

The corresponding solutions $\vec{E}_j^+$, $\vec{E}_j^-$, and $\vec{E}_j^=$ of the Maxwell equations can be formulated according to the equations (see [3])

$$E_j^+ = \sqrt{\frac{\varepsilon_j}{\mu_j}}(\Lambda_{j1}^+)^m(\Lambda_{j2}^+)^n\left\{(\det\gamma_j^+)^{\frac{1}{2}}\exp\left[\frac{i}{2v}\left(\frac{d\gamma_j^+}{ds}(\gamma_j^+)^{-1}\vec{x},\vec{x}\right) + i\omega\int_{e_{j-1}}^{s_j}\frac{ds}{v(s)}\right]\right\}\vec{C}_j^{\,+}, \tag{3.14}$$

$$\vec{E}_j^- = \sqrt{\frac{\varepsilon_j}{\mu_j}}(\Lambda_{j1}^-)^m(\Lambda_{j2}^-)^n\left\{(\det\gamma_j^-)^{\frac{1}{2}}\exp\left[\frac{-i}{2v}\left(\frac{d\gamma_j^-}{ds}(\gamma_j^-)^{-1}\vec{x},\vec{x}\right) - i\omega\int_{e_j}^{s_j}\frac{ds}{v(s)}\right]\right\}\vec{C}_j \tag{3.15}$$

$$\vec{E}_j^= = \sqrt{\frac{\varepsilon_j}{\mu_j}}(\Lambda_{j1}^=)^m(\Lambda_{j2}^=)^n\left\{(\det\gamma_j^=)^{\frac{1}{2}}\exp\left[\frac{i}{2v}\left(\frac{d\gamma_j^=}{ds}(\gamma_j^=)^{-1}\vec{x},\vec{x}\right) - i\omega\int_{e_j}^{s_j}\frac{ds}{v(s)}\right]\right\}\vec{C}_j^{\,=}, \tag{3.16}$$

where $\vec{C}_j = (C_{j\xi}, C_{j\eta})$, $C_{j\xi}, C_{j\eta}$ are arbitrary constants and γ_j^+, γ_j^-, and $\gamma_j^=$ are matrices whose columns are the vectors $\vec{\gamma}_{jk}$, for example, $\gamma_j^+ = \|\vec{\gamma}_{j1}^{\,+}, \vec{\gamma}_{j2}^{\,+}\|$ [the four linearly independent solutions γ_{j1}, γ_{j2}, γ_{j1}^*, and γ_{j2}^* of Eqs. (1.5) are normalized as follows (see [2, 3, 6])]:

$$\left.\begin{aligned} &\left(\frac{d\vec{\gamma}_{jk}}{ds_j}, \vec{\gamma}_{je}\right) - \left(\vec{\gamma}_{jk}, \frac{d\vec{\gamma}_{je}}{ds_j}\right) = iv_j(s_j)\delta_{ke}, \quad k = 1, 2, \quad e = 1, 2, \\ &\left(\frac{d\vec{\gamma}_{jk}}{ds_j}, \vec{\gamma}_{je}^{\,*}\right) - \left(\gamma_{jk}, \frac{d\gamma_{je}^*}{ds_j}\right) = 0 \qquad \delta_{k,e} = \begin{cases}1, & k = e\\ 0, & k \neq e\end{cases}\end{aligned}\right\} \tag{3.17}$$

The operators Λ_{jk} have the form

$$\Lambda_{jk}=\frac{1}{i}\left(\vec{\gamma}_{jk}^{*},\nabla_j\right)-\frac{1}{v_j}\left(\frac{d\gamma_{jk}^{*}}{ds_j},\vec{x}_j\right),$$

$$\vec{x}_j=(x_j,y_j),\quad x_j=\sqrt{\omega}\,\xi_j,\quad y_j=\sqrt{\omega}\,\eta_j,\quad \nabla_j=\left(\frac{\partial}{\partial x_j},\frac{\partial}{\partial y_j}\right),\quad (\vec{f},\vec{g})=\sum_{k=1}^{2}f_k g_k^{*}, \tag{3.18}$$

the asterisk (*) denoting the complex conjugate.

The vector $\vec{H}$ is determined from Eqs. (3.6)-(3.10).

The fields $\vec{E}$ and $\vec{H}$ must be uniquely determined in each partial resonator by equations of the type (3.14)-(3.16). The component $\vec{E}_j^{-}$ characterizes a "downward"-traveling wave, $\vec{E}_j^{=}$ characterizes the same kind of wave, and, therefore, the uniqueness requirement implies that $\vec{E}_j^{=}=\vec{E}_j^{-}$. Hence, taking (3.13) into account, along with the properties of the operators Λ (see [2]) and the requirement (3.4), we obtain the following condition for arbitrary vectors $\vec{C}_j^{=}$ and $\vec{C}_j^{-}$

$$\vec{C}_j^{=}=\vec{C}_j^{-}\exp\left[-\frac{\alpha_1^{(j)}(2m+1)+\alpha_2^{(j)}(2n+1)}{2}\right]. \tag{3.19}$$

The formulated vectors $\vec{E}$ and $\vec{H}$ must satisfy the boundary conditions (3.2) and (3.3).

The action of the operators $(\Lambda_{j1})^m\ (\Lambda_{j2})^n$ on the expression in the braces in (3.14)-(3.16) is equivalent to the multiplication of that expression by polynomials $\mathcal{P}_{m+n}(s;x,y)$ of degree $m+n$ in x and y with coefficients depending on s [2]). For $s=\ell_j$, $j=0,1,\ldots,N$, the boundary conditions (3.2) and (3.3) reduce to the following:

1) equality of the polynomials $(\det\gamma)^{\frac{1}{2}}\cdot\mathcal{P}_{m+n}(s;x,y)$ in the exponentials;

2) equality of the quadratic forms in the exponents;

3) the following relations for $N=3$:

$$\vec{C}_1^{+}+\vec{C}_1^{-}e^{-i\omega\int_{\ell_1}^{\ell_0}\frac{ds}{v_1}}=0$$

$$\sqrt[4]{\frac{\varepsilon_j}{\mu_j}}\Bigg|_{\ell_j}\left[\vec{C}_j^{+}e^{i\omega\int_{\ell_{j-1}}^{\ell_j}\frac{ds}{v_j}}+\vec{C}_j^{=}\right]=\sqrt[4]{\frac{\varepsilon_{j+1}}{\mu_{j+1}}}\Bigg|\left[\vec{C}_{j+1}^{-}e^{-i\omega\int_{\ell_{j+1}}^{\ell_j}\frac{ds}{v_{j+1}}}+\vec{C}_{j+1}^{+}\right], \tag{3.20}$$

$$\sqrt[4]{\frac{\mu_j}{\varepsilon_j}}\Bigg|_{\ell_j}\left[\vec{C}_j^{+}e^{i\omega\int_{\ell_{j-1}}^{\ell_j}\frac{ds}{v_j}}+\gamma^{-1}\vec{C}_j^{=}\right]=\sqrt[4]{\frac{\mu_{j+1}}{\varepsilon_{j+1}}}\Bigg|_{\ell_j}\left[\gamma^{-1}\vec{C}_{j+1}^{-}e^{-i\omega\int_{\ell_{j+1}}^{\ell_j}\frac{ds}{v_{j+1}}}+\gamma\vec{C}_{j+1}^{+}\right],\quad j=1,2,\quad \gamma=\begin{pmatrix}0&-1\\1&0\end{pmatrix}, \tag{3.21}$$

$$\vec{C}_3^{+}e^{i\omega\int_{\ell_2}^{\ell_3}\frac{ds}{v_3}}+\vec{C}_3^{=}=0. \tag{3.22}$$

We shall analyze the equality conditions 1) and 2) in detail in the next section, assuming for the time being that they are fulfilled.

Relations (3.19)-(3.22) represent a homogeneous system of linear algebraic equations in the arbitrary constant, which has a nontrivial solution if the following condition is met:

$$(\rho_1+\rho_2)|_{e_1}(\rho_2+\rho_3)|_{e_2}\sin(\nu_1+\nu_2+\nu_3)+(\rho_2-\rho_1)|_{e_1}(\rho_2+\rho_3)|_{e_2}\sin(\nu_1-\nu_2-\nu_3)+$$
$$+(\rho_2-\rho_1)|_{e_1}(\rho_3-\rho_2)|_{e_2}\sin(\nu_1-\nu_2+\nu_3)+(\rho_1+\rho_2)|_{e_1}(\rho_3-\rho_2)|_{e_2}\sin(\nu_1+\nu_2-\nu_3)=0 \tag{3.23}$$

in which

$$\rho_j=\sqrt{\frac{\mu_j}{\varepsilon_j}}, \quad \nu_j=\frac{\alpha_1^{(j)}(2m+1)+\alpha_2^{(j)}(2n+1)}{2}+\omega\int_{e_{j-1}}^{e_j}\frac{ds}{v_j}.$$

Relation (3.23) is the equation for the frequencies ω_{mn} of a three-layer resonator (cf. the corresponding equation in [1]).

The solutions ω_{mn} of Eq. (3.23) (and only they) are eigenvalues of the self-adjoint problem

$$\frac{1}{i}\frac{\partial W_j^+}{\partial y}=\frac{\omega}{v(y)}W_j^+,$$
$$-\frac{1}{i}\frac{\partial W_j^-}{\partial y}+\frac{\sum_{\kappa=1}^{2}\alpha_\kappa^{(j)}\left(\frac{1}{2}+q_\kappa\right)}{e_j-e_{j-1}}W_j^-=\frac{\omega}{v(y)}W_j^- \quad (q_1=m,\ q_2=n),$$
$$(W_1^++W_1^-)|_{e_0}=0,$$
$$(W_j^++W_j^-)|_{e_j}=(W_{j+1}^++W_{j+1}^-)|_{e_j}, \quad j=1,2,$$
$$\rho_j(W_j^+-W_j^-)|_{e_j}=\rho_{j+1}(W_{j+1}^+-W_{j+1}^-)|_{e_j}, \quad j=1,2,$$
$$(W_3^++W_3^-)|_{e_3}=0.$$

It follows from the latter that Eq. (3.23) has an innumerable set of roots ω_{mn}, all of which are real.

§ 4. Compatibility Conditions

We now investigate the problem of when the boundary conditions (3.2) and (3.3) are satisfied.*

We first consider the zeroth (longitudinal) mode: $m=n=0$.

Relations (3.2) have the following form for the fields $\vec{E}$ and $\vec{H}$ formulated above, taking into account their transversality in the incident and reflected waves and the smallness of the angle of ray incidence on the interface:

$$(\vec{E}_{j+1}^-+\vec{E}_{j+1}^+)|_{S_j}=(\vec{E}_j^++\vec{E}_j^=)|_{S_j},$$
$$(\vec{H}_{j+1}^-+\vec{H}_{j+1}^+)|_{S_j}=(\vec{H}_j^++\vec{H}_j^=)|_{S_j}, \tag{4.1}$$

where $\vec{E}^-$, $\vec{E}^+$, and $\vec{E}^=$ are determined by Eqs. (3.14)-(3.16) and $\vec{H}^-$, $\vec{H}^+$, and $\vec{H}^=$ are expressed in terms thereof according to (3.6)-(3.10).

*Up to terms of order $\omega^{-\frac{1}{2}}$ in the neighborhood of the point of incidence.

Conditions (4.1) are fulfilled if:

1)

$$\det \gamma_j^+ |_{e_j} = \det \gamma_j^= |_{e_j} = \det \gamma_{j+1}^+ |_{e_j} = \det \gamma_{j+1}^- |_{e_j}; \tag{4.2}$$

2)

$$\left\{\frac{1}{v_j}\left[(\gamma_j^+)'(\gamma_j^+)^{-1} + 2D_j\right]\right\}_{e_j} = \left\{-\frac{1}{v_j}\left[(\gamma_j^=)'(\gamma_j^=)^{-1} + 2D_j\right]\right\}_{e_j} =$$

$$= \left\{\frac{1}{v_{j+1}}\left[(\gamma_{j+1}^+)'(\gamma_{j+1}^+)^{-1} + 2D_j\right]\right\}_{e_j} = \left\{-\frac{1}{v_{j+1}}\left[(\gamma_{j+1}^-)'(\gamma_{j+1}^-)^{-1} + 2D_j\right]\right\}_{e_j}, \tag{4.3}$$

where the prime denotes differentiation with respect to s [Eqs. (4.3) are derived as follows: we regard the upper integration limit in (3.14)-(3.16) as a point on the surface S_j and expand the function $\int_{s_0}^{s_j(\xi_j,\eta_j)} \frac{ds_j}{v(s_j,0,0)}$ in a series in a neighborhood of the point of incidence $s_j = e_j$, $\xi_j = \eta_j = 0$; then the equality condition on the quadratic forms in the exponents takes the form (4.3)];

3) conditions (3.21) are fulfilled.

It follows from (3.12) and (3.13) (i.e., from the stability of each partial resonator and from the fact that the matrices γ_j are formed from the Floquet solutions for the jth resonator) that conditions (3.3) are fulfilled at the boundaries S_0 and S_N, while at the boundaries S_j between the media, the matrices γ_j satisfy the relations

$$\begin{aligned} &\operatorname{Re} G_j^+ |_{e_j} = \operatorname{Re} G_j^= |_{e_j} = -2D_j, \quad j = 1, 2, \ldots, N-1, \\ &\operatorname{Im} G_j^+ |_{e_j} = -\operatorname{Im} G_j^= |_{e_j}, \\ &\operatorname{Re} G_{j+1}^+ |_{e_j} = \operatorname{Re} G_{j+1}^- |_{e_j} = -2D_j, \\ &\operatorname{Im} G_{j+1}^+ |_{e_j} = -\operatorname{Im} G_{j+1}^- |_{e_j}, \end{aligned} \tag{4.4}$$

where $G_j \equiv \left(\frac{d}{ds}\gamma_j\right)(\gamma_j)^{-1}$.

Equations (4.3), taking (4.4) into account, are fulfilled if

$$\frac{1}{v_j} \operatorname{Im} G_j^= \Big|_{e_j} = \frac{1}{v_{j+1}} \operatorname{Im} G_{j+1}^- \Big|_{e_j}, \tag{4.5}$$

where the following relation holds for the matrices G_j (see [2]):

$$G_j^- |_{e_j} = G_j^= |_{e_j}. \tag{4.6}$$

It is readily deduced from the normalization conditions (3.17) (see [2]) that

$$\operatorname{Im} G_j = \frac{1}{2} v_j \left[\gamma_j \gamma_j^*\right]^{-1} \tag{4.7}$$

(the asterisk * denoting the Hermitian conjugate).

Equation (4.5) reduces to the form

$$\gamma_j^- \gamma_j^{-*} |_{e_j} = \gamma_{j+1}^- \gamma_{j+1}^{-*} |_{e_j}, \quad \gamma_j^- = \|\vec{\gamma}_{j1}^{\,*}, \vec{\gamma}_{j2}^{\,*}\|. \tag{4.8}$$

The matrices $\gamma_j^- \gamma_j^{-*}$ are symmetric, real, and of second order, so that relation (4.8) essentially represents three conditions, depending on the parameters of the medium, on the components of the vectors $\vec{\gamma}_{jk}^*$ and $\vec{\gamma}_{(j+1)k}^*$, $k=1,2$.

It follows from (4.8) that the matrices $\gamma_j^-(c_j)$ and $\gamma_{j+1}^-(c_j)$ differ only by a unitary transformation, so that $\det \gamma_j^=(c_j) = \det \gamma_{j+1}^-(c_j)$ [the equality conditions $\det \gamma_j^=(e_j) = \det \gamma_j^+(e_j)$ and $\det \gamma_{j+1}^+(e_j) = \det \gamma_{j+1}^-(e_j)$ follow from (3.12) and (3.13)].

Using the properties of the operators $(\Lambda_{j1})^m$ and $(\Lambda_{j2})^n$ as in [2], we can easily show that the equality of the corresponding polynomials in the exponentials reduces to relations (4.4)-(4.8). Consequently, when the three additional conditions (4.8) on the Floquet solutions are fulfilled, the boundary conditions (3.2) and (3.3) are satisfied for any mode, and the eigenfrequencies of the resonator are approximately determined from Eqs. (3.23).

It follows from Eqs. (4.4)-(4.8) that the vectors $\vec{\gamma}_{jk}(c_j)$ and $\vec{\gamma}_{(j+1)k}(c_j)$ differ only by a unitary transformation, whereupon we readily infer that the eigenvectors $\vec{a}_j^-$ and $\vec{a}_{j+1}^-$ of the matrices $\Gamma_{j+1}^{B_j} \Gamma_j^{B_j}$ and $\Gamma_j \Gamma_{j-1}$ differ by a unitary transformation and, hence, the matrices $\Gamma_{j+1}^{B_j} \Gamma_j^{B_j}$ and $\Gamma_j \Gamma_{j-1}$ are orthogonalized by one and the same matrix. Inasmuch as this result holds for any j, we conclude that all the matrices Φ from § 2 are orthogonalized by one and the same transformation.

On the other hand, it follows from the stability of the system of rays in the multilayered resonator, i.e., from the fact that all the matrices Φ are orthogonalized by the same matrix, that:

1) the eigenvalues for any matrix $\Gamma_j \Gamma_{j-1}$, $j=1,\ldots,N$, are equal to unity in the modulus and the proper divisors are prime (stability of each partial resonator);

2) the eigenvectors of the matrices $\Gamma_j \Gamma_{j-1}$ and $\Gamma_{j+1}^{B_j} \Gamma_j^{B_j}$ differ only by a unitary transformation [it is easily demonstrated that in this event the Floquet solutions at the layer interfaces satisfy the compatibility conditions (4.8)].

We deduce the following result on the basis of the entire foregoing discussion: In order for the boundary-value problem (3.1)-(3.3) to be solvable by the parabolic equation method it is necessary and sufficient that the ray system in each partial resonator be stable under the compatibility conditions (4.8).

§ 5. Compatibility Conditions in the Case of Unfolded Mirrors (for Homogeneous Layers)

When the medium in each jth layer is homogeneous, it is a simple matter to determine the eigenvalues of the matrices $\Gamma_j \Gamma_{j-1}$ (see [7]), to formulate the eigenvectors $\vec{a}_j^-$ and Floquet solutions $\vec{\gamma}_{jk}^-(s_j)$, and to write conditions (4.8) specifically in the form of three solutions between the resonator parameters.

We now concern ourselves with the execution of this program.

The resonator axis is a straight line in the case of homogeneous media. We arbitrarily pick out two adjoining media. We choose the coordinate system (s, ξ, η) so that the directions of the unit vectors $\vec{e}_\xi$, and $\vec{e}_\eta$ on the boundary S_j between the two media will coincide with the principal directions of curvature of that surface. We place the reference origin for the length s at the point of intersection of the axis with the boundary S_j between the media (Fig. 2). Then the unit vectors $\vec{e}_\xi$ and $\vec{e}_\eta$ will be rotated through an angle θ_j relative to the principal directions of the surface S_{j-1} at $s_j = -e_j$, through an angle θ_{j+1} relative to the principal directions of S_{j+1} at $s_j = \ell_{j+1}$.

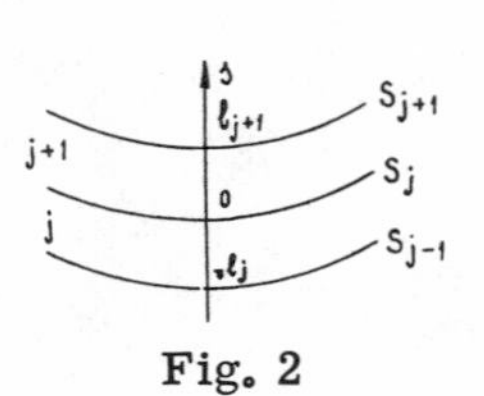

Fig. 2

We rotate the Cartesian coordinates systems $(\varsigma_{1\ell}, \varsigma_{2\ell}, \varsigma_{3\ell})$, $\ell = j-1, j, j+1$, on each surface so that $u(\theta_\ell) = I$ in (1.7). Then the elements of the quadratic forms (1.3) $\mathcal{D}_{ik}^{(\ell)}$, $\ell = j-1, j, j+1$, have the following form for the surface S_ℓ:

$$\mathcal{D}_{11}^{(\ell)} = \frac{1}{2}\left(\frac{\cos^2\tilde{\theta}_\ell}{r_1^{(\ell)}} + \frac{\sin^2\tilde{\theta}_\ell}{r_2^{(\ell)}}\right), \tag{5.1}$$

$$\mathcal{D}_{12}^{(\ell)} = \pm\frac{1}{2}\sin\tilde{\theta}_\ell\cos\tilde{\theta}_\ell\left(\frac{1}{r_1^{(\ell)}} - \frac{1}{r_2^{(\ell)}}\right), \tag{5.2}$$

$$\mathcal{D}_{22}^{(\ell)} = \frac{1}{2}\left(\frac{\sin^2\tilde{\theta}_\ell}{r_1^{(\ell)}} + \frac{\cos^2\tilde{\theta}_\ell}{r_2^{(\ell)}}\right), \tag{5.3}$$

where $r_1^{(\ell)}$, and $r_2^{(\ell)}$ are the principal radii of curvature of the surface S_ℓ, $\tilde{\theta}_{j-1} \equiv \theta_j$, $\tilde{\theta}_j \equiv 0$, and $\tilde{\theta}_{j+1} \equiv \theta_{j+1}$.

Equations (1.5) have the following form for homogeneous media:

$$\vec{\gamma}_j'' = 0. \tag{5.4}$$

In each layer we choose a basis (see §1) as follows:

$$\vec{\gamma}_1^{(\ell)} = \begin{pmatrix}1\\0\end{pmatrix},$$
$$\vec{\gamma}_2^{(\ell)} = \begin{pmatrix}s\\0\end{pmatrix},$$
$$\vec{\gamma}_3^{(\ell)} = \begin{pmatrix}0\\1\end{pmatrix},$$
$$\vec{\gamma}_4^{(\ell)} = \begin{pmatrix}0\\s\end{pmatrix},$$

whereupon the matrix

$$w(s) = \begin{pmatrix} 1 & s & 0 & 0 \\ 0 & 1 & 0 & 0 \\ 0 & 0 & 1 & s \\ 0 & 0 & 0 & 1 \end{pmatrix} \tag{5.5}$$

and the following relations hold:

$$w^{-1}(s) = w(-s), \tag{5.6}$$

$$w(0) = I. \tag{5.7}$$

Using Eqs. (5.1)-(5.7) we construct the matrices $\Gamma_j\Gamma_{j-1}$ and $\Gamma_{j+1}\Gamma_j$ according to Eqs. (1.7) and find their eigenvalues, i.e., we solve the equations

$$\det\left(\Gamma_j\Gamma_{j-1} - \lambda^{(j)} I\right) = 0 \tag{5.8}$$

$$\det\left(\Gamma_{j+1}\Gamma_j - \lambda^{(j+1)} I\right) = 0. \tag{5.9}$$

We construct the matrices $\gamma_\ell^-(\gamma_\ell^-)^*$, $\ell = j, j+1$ for the case in which the directions of principal curvature of the surfaces S_j, S_{j+1} and S_{j-1}, are rotated relative to one another $(\theta_j \neq 0, \theta_{j+1} \neq 0)$. We assume that the stability conditions for the ℓth resonator (see [7]) are fulfilled and that the roots of Eqs. (5.8) and (5.9) are pairwise complex conjugates and equal in the modulus. For the numbers $\lambda_\kappa^{(\ell)}$, $\kappa = 1, 2, 3, 4$, we find the eigenvectors $\vec{a}_\kappa^{(\ell)} = (a_{\kappa 1}, a_{\kappa 2}, a_{\kappa 3}, a_{\kappa 4})$ of the matrices $\Gamma_\ell \Gamma_{\ell-1}$, $\ell = j, j+1$, up to an arbitrary constant multiplier $c_\kappa^{(\ell)}$. Conditions (4.8) involve the products $\tilde{c}_\kappa^{(\ell)} = c_\kappa^{(\ell)} (c_\kappa^{(\ell)})^*$, which we determine from the normalization conditions (3.17). The Floquet solutions $\vec{\gamma}_{\ell\kappa}^{-}(s)$ are expressed in terms of the vectors $\vec{a}_\kappa^{(\ell)}$, $\kappa = 1, 2, 3, 4$, as follows:

$$\vec{\gamma}_{\ell\kappa}^{-}(s) = \begin{pmatrix} a_{\kappa 1}^{(\ell)} + a_{\kappa 2}^{(\ell)} s \\ a_{\kappa 3}^{(\ell)} + a_{\kappa 4}^{(\ell)} s \end{pmatrix}. \tag{5.10}$$

In our chosen coordinate system condition (4.8) must hold for $s = 0$ yielding the following three equations:

1)
$$\sum_{\kappa=1}^{2} a_{\kappa 1}^{(j)} a_{\kappa 1}^{(j)*} = \sum_{\kappa=1}^{2} a_{\kappa 1}^{(j+1)} a_{\kappa 1}^{(j+1)*}; \tag{5.11}$$

2)
$$\sum_{\kappa=1}^{2} a_{\kappa 3}^{(j)} a_{\kappa 3}^{(j)*} = \sum_{\kappa=1}^{2} a_{\kappa 3}^{(j+1)} a_{\kappa 3}^{(j+1)*}; \tag{5.12}$$

3)
$$\sum_{\kappa=1}^{2} a_{\kappa 1}^{(j)} a_{\kappa 3}^{(j)*} = \sum_{\kappa=1}^{2} a_{\kappa 1}^{(j+1)} a_{\kappa 3}^{(j+1)*}. \tag{5.13}$$

Hence, substituting the specific expressions for $a_{\kappa q}^{(\ell)}$, we obtain the three compatibility conditions

1)
$$\frac{1+\frac{\ell_j}{\tau_2^{(j)}}}{b_1^{(j)} - 2(\varkappa_1^{(j)} + \Omega_1^{(j)})} B_1^{(j)} - \frac{1+\frac{\ell_j}{\tau_2^{(j)}}}{b_2^{(j)} - 2(\varkappa_1^{(j)} + \Omega_1^{(j)})} B_2^{(j)} = \frac{1-\frac{\ell_{j+1}}{\tau_2^{(j)}}}{b_1^{(j+1)} - 2(\varkappa_1^{(j+1)} + \Omega_1^{(j+1)})} B_1^{(j+1)} - \frac{1-\frac{\ell_{j+1}}{\tau_2^{(j)}}}{b_2^{(j+1)} - 2(\varkappa_1^{(j+1)} + \Omega_1^{(j+1)})} B_2^{(j+1)}; \tag{5.14}$$

2)
$$\frac{b_1^{(j)} - 2(\varkappa_1^{(j)} + \Omega_1^{(j)})}{1+\frac{\ell_j}{\tau_2^{(j)}}} B_1^{(j)} - \frac{b_2^{(j)} - 2(\varkappa_1^{(j)} + \Omega_1^{(j)})}{1+\frac{\ell_j}{\tau_2^{(j)}}} B_2^{(j)} = \frac{b_1^{(j+1)} - 2(\varkappa_1^{(j+1)} + \Omega_1^{(j+1)})}{1-\frac{\ell_{j+1}}{\tau_2^{(j)}}} B_1^{(j+1)} - \frac{b_2^{(j+1)} - 2(\varkappa_1^{(j+1)} + \Omega_1^{(j+1)})}{1-\frac{\ell_{j+1}}{\tau_2^{(j)}}} B_2^{(j+1)}; \tag{5.15}$$

3)
$$B_1^{(j)} - B_2^{(j)} = B_1^{(j+1)} - B_2^{(j+1)}, \tag{5.16}$$

in which the following notation is used:

$$B_\kappa^{(j)} \equiv \frac{v_j \ell_j}{b_1^{(j)} - b_2^{(j)}}, \quad \kappa = 1, 2, \quad v_j \equiv v_j(\ell_j), \quad v_{j+1} \equiv v_{j+1}(\ell_j),$$

$v_j(s)$ is the electromagnetic wave propagation velocity in the jth medium,

$$b_{1,2}^{(j)} = \cos\alpha_{1,2}^{(j)} = \varkappa_1^{(j)} + \varkappa_2^{(j)} - \Omega_1^{(j)} - \Omega_2^{(j)} \pm (\varkappa_1^{(j)} - \varkappa_2^{(j)}) \sqrt{1 - (\Omega_1^{(j)} + \Omega_2^{(j)}) \frac{2(\varkappa_1 + \varkappa_2) - \Omega_1 - \Omega_2 - 4}{(\varkappa_1 - \varkappa_2)^2}}$$

($\alpha_1^{(j)}$ and $\alpha_2^{(j)}$ are the Floquet exponents in the jth medium), and

$$\varkappa_K^{(j)} = 1 + 2\ell_j\left(\frac{1}{\tau_K^{(j)}} - \frac{1}{\tau_K^{(j-1)}}\right) - \frac{2\ell_j^2}{\tau_K^{(j-1)}\tau_K^{(j)}}, \quad K = 1, 2,$$

$$\Omega_K^{(j)} = 2\ell_j\left(1 + \frac{\ell_j}{\tau_K^{(j)}}\right)\left(\frac{1}{\tau_K^{(j)}} - \frac{1}{\tau_\ell^{(j)}}\right)\sin^2\theta_j, \quad K=1,2,\ \ell=1,2,\ \ell \neq K.$$

In the special case in which $\theta_j = \theta_{j+1} = 0$ the matrices Γ_{j-1}, Γ_j and Γ_{j+1} are quasi diagonal, the problem is partitioned into two two-dimensional problems,

$$\vec{a}_1^{(j)} = \left(a_{11}^{(j)},\ a_{12}^{(j)},\ 0,\ 0\right),$$
$$\vec{a}_2^{(j)} = \left(a_{11}^{(j)*},\ a_{12}^{(j)*},\ 0,\ 0\right),$$
$$\vec{a}_3^{(j)} = \left(0,\ 0,\ a_{33}^{(j)},\ a_{34}^{(j)}\right),$$
$$\vec{a}_4^{(j)} = \left(0,\ 0,\ a_{33}^{(j)*},\ a_{34}^{(j)*}\right),$$

Eq. (5.13) is the equation 0 = 0, and only the two compatibility conditions

$$v_j\sqrt{\frac{1 - \frac{\tau_K^{(j-1)}}{\ell_j}}{\left(1 + \frac{\tau_K^{(j)}}{\ell_j}\right)\left(\frac{\tau_K^{(j-1)}}{\ell_j} - \frac{\tau_K^{(j)}}{\ell_j} - 1\right)}} = v_{j+1}\sqrt{\frac{\frac{\tau_K^{(j+1)}}{\ell_{j+1}} + 1}{\left(1 - \frac{\tau_K^{(j)}}{\ell_j}\right)\left(\frac{\tau_K^{(j)}}{\ell_{j+1}} - \frac{\tau_K^{(j+1)}}{\ell_{j+1}} - 1\right)}}, \quad K=1,2,$$

remain, which exactly coincide with the compatibility conditions derived in [1] for the two-dimensional case.

LITERATURE CITED

1. Babich, V. M., Eigenmodes of a multilayered resonator, Seminars in Mathematics, Vol. 15, Consultants Bureau, New York (1971), pp. 1-22.
2. Popov, M. M., Eigenmodes of multimirror resonators, Vestnik Leningrad. Univ. (LGU), No. 22 (1969).
3. Pankratova, T. F., Eigenmodes in an annular resonator; the vectoral problem, Seminars in Mathematics, Vol. 15, Consultants Bureau, New York (1971), pp. 67-79.
4. Cassels, J. W. S., An Introduction to Diophantine Approximations, Cambridge Univ. Press (1957).
5. Smirnov, V. I., A Course of Higher Mathematics, Pergamon, Oxford-New York (1964).
6. Babich, V. M., Eigenfunctions concentrated in a neighborhood of a closed geodesic, Seminars in Mathematics, Vol. 9, Consultants Bureau, New York (1970), pp. 7-26.
7. Popov, M. M., Resonators for lasers with unfolded directions of principal curvatures, Opt. i Spektrosk., 25(3):394-400 (1968).
8. Vilenkin, N. Ya., Special Functions and the Theory of Group Representations, Nauka, Moscow (1965).

SHADOW ZONE AT THE BOUNDARY OF AN INHOMOGENEOUS HALF-SPACE

N. V. Tsepelev

Consider an inhomogeneous half-space $z>0$ in which wave propagation processes are described by the wave equation and the propagation velocity is such that the following expansion is applicable for small z:

$$v(z)=v(0)-v_1(0)\,z^{\alpha}+\cdots, \quad \alpha>0, \tag{1}$$

where $v_1(0)>0$. It can be shown that if $\alpha\in(0,2)$ a geometric shadow zone will be formed in the vicinity of the boundary of the half-space relative to a source placed sufficiently close to the boundary. We propose to investigate the field in that zone.

Let a stationary source of oscillations be situated at a point M_0 with coordinates $x=0$, $z=H$. It is required to analyze the wave field in the shadow zone at high frequencies. We shall assume for definiteness that the boundary of the half-space is rigidly attached, as the solution of other problems with different boundary conditions is analogous. Let

$$u=U(x,z,0,H)\,e^{-i\omega t}. \tag{2}$$

The function $U(x,z,0,H)$ must satisfy the equation

$$\Delta U+\frac{\omega^2}{v^2(z)}U=-\delta(x)\,\delta(z-H), \tag{3}$$

in which δ is the Dirac delta function, as well as the boundary condition

$$U\big|_{z=0}=0. \tag{4}$$

Also, let the absorption limit principle hold as $(x^2+z^2)\to\infty$.

The ray pattern near the boundary $z=0$ will be as shown in Fig. 1. To the right of the point $x=a$ is the geometric shadow zone. We shall investigate the wave field therein by the parabolic equation method. Accordingly, we introduce the attenuation function

$$V(x,z)=U(x,z,0,H)\,e^{-i\kappa x}, \tag{5}$$

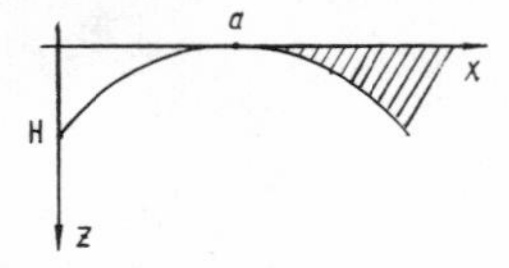

Fig. 1

in which $\kappa=\frac{\omega}{v(0)}$, and we initially consider the solutions $U(x,z)$ of Eq. (3) with zero on the right-hand side. We then obtain for $V(x,z)$

$$\frac{\partial^2 V}{\partial x^2}+\frac{\partial^2 V}{\partial z^2}+2i\kappa\frac{\partial V}{\partial x}+\kappa^2\left[\frac{v^2(0)}{v^2(z)}-1\right]V=0. \tag{6}$$

To the right of the point a (Fig. 1) we consider the layer contiguous with the x axis, assuming its thickness is of the order $O(\kappa^{-\frac{2}{2+\alpha}})$,* and conduct our ensuing analysis for z satisfying the inequality

$$|z| < A\kappa^{-\frac{2}{2+\alpha}}, \quad A = \text{const.} \tag{7}$$

In accordance with (1) we expand the function $v^{-2}(z)$ in this layer in a power series on z^{α}:

$$\frac{v^2(0)}{v^2(z)} = 1 + 2\varepsilon z^{\alpha} + \dots, \tag{8}$$

in which

$$\varepsilon = \frac{v_1(0)}{v(0)} > 0, \tag{9}$$

and we introduce the new coordinates

$$\begin{aligned} \zeta &= (2\varepsilon\kappa^2)^{\frac{1}{2+\alpha}} z, \\ \eta &= (2\varepsilon\kappa)^{\frac{2}{2+\alpha}} \kappa^{-\frac{\alpha}{2+\alpha}} \frac{x}{2}. \end{aligned} \tag{10}$$

The new scale on z is based on the consideration that the coordinate ζ in the layer (7) remain bounded as $\kappa \to \infty$, while the scale on x is based on the examination of a grazing ray in the layer. Then, up to terms of order $O(\kappa^{-\frac{2-\alpha}{2+\alpha}})$, Eq. (6) assumes the form

$$\frac{\partial^2 V}{\partial \zeta^2} + \zeta^{\alpha} V + i\frac{\partial V}{\partial \eta} = 0. \tag{11}$$

Its solution is the function

$$V = A e^{ip\eta} F_{\alpha}(\zeta, p), \tag{12}$$

where A and p are arbitrary constants and F_{α} satisfies the equation

$$F_{\alpha}'' + (\zeta^{\alpha} - p) F_{\alpha} = 0. \tag{13}$$

When $\alpha = 1$ (this case has been thoroughly analyzed in [1]), F_{α} is an Airy function.

For the discussion that follows we need to establish certain properties of the solutions of Eq. (13). We first explicate their behavior as $\zeta \to \infty$, as well as the positions of the roots of the equation $F_{\alpha}(\zeta_0, p) = 0$ on the complex plane of p for fixed ζ_0. The first of these properties is required in order to satisfy the absorption limit principle, while the second is essential to the fulfillment of the boundary condition (4).

In order to answer the stated questions regarding the behavior of the solutions it suffices to formulate their uniform asymptotic behavior as $|p| \to \infty$ in a neighborhood of the single turning point of Eq. (13). We introduce in the latter the new independent variable

$$\mathfrak{z} = p^{-1/\alpha} \zeta. \tag{14}$$

Inasmuch as we have not insisted that the parameter p be real, we shall investigate the solutions of the transformed equation

*The choice of layer thickness is based on the condition that the second term of Eq. (6) and the principal term of the last term in the expansion on z are identical in order of magnitude in the layer.

$$F_\alpha'' + p^{\frac{2+\alpha}{\alpha}} (s^\alpha - 1) F_\alpha = 0 \tag{15}$$

in the complex plane of s. In order to make the proper choice of branch of the many-valued function s^α we draw a branch cut from the origin in the direction of $-\infty$ and fix it by the condition that $s^\alpha > 0$ when $s > 0$.

We formulate the solution of Eq. (15) by the standard-equation method, choosing the following as the standard:

$$F'' + x^\alpha F = 0, \qquad 0 < \alpha < 2, \tag{16}$$

where

$$F = \sqrt{x}\, H^{(\nu)}_{\frac{1}{2+\alpha}} \left(\frac{2}{2+\alpha} x^{\frac{2+\alpha}{2}} \right), \tag{17}$$

in which $H^{(\nu)}_{\frac{1}{2+\alpha}}$ is a Hankel function. Using the solution (17) for (15), we proceed in the usual way to obtain

$$F_\alpha = \frac{\sqrt{I(s)}}{(s^\alpha - 1)^{1/4}} H^{(\nu)}_{\frac{1}{2+\alpha}} \left(\pm p^{\frac{2+\alpha}{2\alpha}} I(s) \right) \left[1 + O\left(p^{-\frac{\alpha+1}{\alpha}} \right) \right], \tag{18}$$

in which

$$I(s) = \int_1^s \sqrt{s^\alpha - 1}\, ds. \tag{19}$$

On the plane of s we draw a branch cut from the point $s = 1$ toward $-\infty$ and fix the branch of the root $\sqrt{s^\alpha - 1}$ by the condition that $\sqrt{s^\alpha - 1} > 0$ when $s > 1$.

We now consider the integral (19) and ascertain its order of growth as $|s| \to \infty$. It can be shown that

$$I(s) = \frac{2}{2+\alpha} s^{\frac{2+\alpha}{2}} \left[1 + O(s^{-\alpha}) \right]. \tag{20}$$

Using this relation and Eq. (14), we write the argument of the Hankel function in (18) as follows when $|s| \to \infty$:

$$\pm p^{\frac{2+\alpha}{2\alpha}} I(s) = \pm \frac{2}{2+\alpha} S^{\frac{2+\alpha}{2}} \left[1 + O(s^{-\alpha}) \right]. \tag{21}$$

The resulting expression makes it possible to choose the sign in the argument of the Hankel function in (18) and to choose the Hankel function itself so as to satisfy the absorption limit principle. Our solution must have the following asymptotic representation as $|p| \to \infty$:

$$F_\alpha = \frac{\sqrt{I(s)}}{(s^\alpha - 1)^{1/4}} H^{(2)}_{\frac{1}{2+\alpha}} \left(-p^{\frac{2+\alpha}{2\alpha}} I(s) \right) \left[1 + O\left(p^{-\frac{\alpha+1}{\alpha}} \right) \right]. \tag{22}$$

Next we consider the positions of the zeros of F_α on the plane of p for fixed s. For the ensuing discussion we need to have $s = 0$ and, on the other hand, to consider the parameter p to be sufficiently large [so as to permit the asymptotic representation of the Hankel function in (22) for a large argument to be used]. Under the given assumptions the roots we seek are approximately determined by the equation

$$e^{2I_0 p^{\frac{2+\alpha}{2\alpha}}} + 2e^{\frac{\pi i}{2}} \cos\frac{\pi}{2+\alpha} = 0, \tag{23}$$

in which

$$I_0 = \int_0^1 \sqrt{1-s^\alpha}\, ds. \tag{24}$$

Let $p_n = \rho_n e^{i\chi_n}$ be the root of Eq. (23), so that

$$\left\{\begin{aligned} \rho_n &= \left[\frac{\ln^2\left(2\cos\frac{\pi}{2+\alpha}\right) + \pi^2\left(2n-\frac{1}{2}\right)^2}{4I_0^2}\right]^{\frac{\alpha}{2+\alpha}}, \\ \operatorname{ctg}\frac{2+\alpha}{2\alpha}\chi_n &= \frac{\ln\left(2\cos\frac{\pi}{2+\alpha}\right)}{\pi\left(2n-\frac{1}{2}\right)}, \end{aligned}\right. \tag{25}$$

$$n = 1, 2, 3, \ldots .$$

In this case χ_n does not exceed $\frac{\pi}{2}$. The accuracy of the resulting equations increases with n. However, even for $n=1$ and not too small α the accuracy is already quite satisfactory, since $\rho_1 \gg 1$.

Using the foregoing results, we can write the sought-after solution of the problem in satisfaction of the absorption limit principle and the boundary condition (4) as follows:

$$U(x,z,0,H) = \sum_{n=1}^{\infty} M_n F_\alpha^{(n)}(\zeta) e^{i\kappa x + p_n \eta}\left[1 + O\left(\kappa^{-\frac{2-\alpha}{2+\alpha}}\right)\right], \tag{26}$$

where

$$F_\alpha^{(n)}(\zeta) = \frac{\sqrt{I_n(\zeta)}}{(\zeta^\alpha - p_n)^{1/4}} H^{(2)}_{\frac{1}{2+\alpha}}(-I_n(\zeta))\left[1 + O\left(p_n^{-\frac{\alpha+1}{\alpha}}\right)\right] \tag{27}$$

and $I_n(\zeta)$ has the form

$$I_n(\zeta) = \int_{p_n^{1/\alpha}}^{\zeta} \sqrt{\zeta^\alpha - p_n}\, d\zeta. \tag{28}$$

In Eq. (26) M_n are arbitrary constants. We determine them by the technique described in [2]. We require that our solution (26) satisfy the inhomogeneous equation (3), i.e., that it represent the Green function for our stated problem. Proceeding from the reciprocity principle, we demand for the Green function that

$$M_n = N_n F_\alpha^{(n)}(\zeta_0), \tag{29}$$

where

$$\zeta_0 = (2\varepsilon \kappa^2)^{\frac{1}{2+\alpha}} H \tag{30}$$

and N_n is independent of the coordinates of either the source or observation point.

Consequently,

$$U(x,z,0,H)=\sum_{n=1}^{\infty} N_n F_\alpha^{(n)}(\varsigma_0)F_\alpha^{(n)}(\varsigma)e^{i[\kappa|x|+p_n|\eta|]}\left[1+O\left(\kappa^{-\frac{2-\alpha}{2+\alpha}}\right)\right]. \tag{31}$$

The latter equation is valid for positive as well as negative x. Inserting it into Eq. (13) and taking the convolution with $\delta(x)$, we obtain, up to principal terms,

$$2i\kappa\sum_{n=1}^{\infty} N_n F_\alpha^{(n)}(\varsigma_0)F_\alpha^{(n)}(\varsigma)=-\delta(z-H)=-(2\varepsilon\kappa^2)^{\frac{1}{2+\alpha}}\delta(\varsigma-\varsigma_0). \tag{32}$$

We now establish an important property of the function $F_\alpha^{(n)}(\varsigma)$ with regard to the ensuing discussion. Using Eq. (13) and the fact that $F_\alpha^{(n)}(0)=0$ and $F_\alpha^{(n)\prime}(\varsigma)\underset{\varsigma\to\infty}{\longrightarrow}0$, we find

$$\int_0^\infty F_\alpha^{(n)}(\varsigma)F_\alpha^{(m)}(\varsigma)\,d\varsigma=\begin{cases}0, & p_m\neq p_n;\\ -[F_\alpha^{(n)\prime}(0)]^2, & p_m=p_n,\end{cases} \tag{33}$$

in which p_ℓ $(\ell=m,n)$ are the roots of the equation $F_\alpha^{(\ell)}(0)=0$.

We multiply Eq. (32) by $F_\alpha^{(m)}(\varsigma)$ and integrate it over ς from 0 to ∞. Using property (33), we obtain

$$N_n=-\frac{i}{2}\left(\frac{2\varepsilon}{\kappa^\alpha}\right)^{\frac{1}{2+\alpha}}\frac{1}{[F_\alpha^{(n)\prime}(0)]^2}. \tag{34}$$

Substituting the resulting expression into (31), we have

$$U=-\frac{i}{2}\left(\frac{2\varepsilon}{\kappa^\alpha}\right)^{\frac{1}{2+\alpha}}\sum_{n=1}^{\infty}\frac{F_\alpha^{(n)}(\varsigma_0)F_\alpha^{(n)}(\varsigma)}{[F_\alpha^{(n)\prime}(0)]^2}e^{i\kappa x+ip_n\eta}\left[1+O\left(\kappa^{-\frac{2-\alpha}{2+\alpha}}\right)\right]. \tag{35}$$

If we take into account the complexity of the expressions for p_n in (25) and bear in mind that $\operatorname{Im} p_n>0$, we can show that only the first term need be retained in the sum (35) for sufficiently large ς_0 and ς, because the contribution of the remaining terms is exponentially small by comparison. Making use in this case of the asymptotic representations of the functions $F_\alpha^{(1)}(\varsigma)$ and $F_\alpha^{(1)}(\varsigma_0)$ and taking the expansion (1) into account, we write the following expression for the field in the shadow zone:

$$U=-\frac{1}{\pi}\left(\frac{2\varepsilon}{\kappa^\alpha}\right)^{\frac{2}{2+\alpha}}\frac{1}{[F_\alpha^{(1)\prime}(0)]^2}\frac{1}{\sqrt{RR_0}}e^{i\omega\tau-\frac{ip_1}{2}(2\varepsilon)^{\frac{2}{2+\alpha}}\kappa^{\frac{2-\alpha}{2+\alpha}}\tilde{x}+\frac{\pi i}{2+\alpha}}\left[1+O\left(\kappa^{-\frac{2-\alpha}{2+\alpha}}\right)\right] \tag{36}$$

Here τ corresponds to the transit time of a propagating disturbance from the source to the point of observation, $\tilde{x}$ is the path length of a ray along the boundary $z=0$ (Fig. 2), and R and R_0 are the geometric divergences on the respective segments BM and HA, where

$$R=\frac{\sqrt{v^2(0)-v^2(z)}}{v(z)},\qquad R_0=\frac{\sqrt{v^2(0)-v^2(H)}}{v(H)}. \tag{37}$$

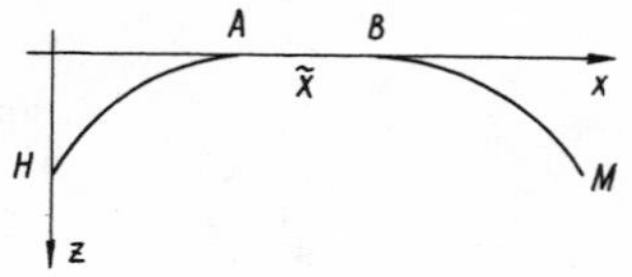

Fig. 2

In the special case of $\alpha=1$, Eq. (36) goes over to the well-known expressions given in [1]. The frequency dependence of the wave amplitude (36) is given by the factor

$$\left(\frac{2\varepsilon}{\kappa^{\alpha}}\right)^{\frac{2}{2+\alpha}} e^{-\frac{\operatorname{Im} \beta_1}{2}(2\varepsilon)^{\frac{2}{2+\alpha}} \kappa^{\frac{2-\alpha}{2+\alpha}} \tilde{x}}, \tag{38}$$

i.e., the disturbance decays exponentially in the shadow zone as $\kappa \to \infty$. The rate of decay depends on α and turns out to be the strongest as $\alpha \to 0$.

LITERATURE CITED

1. Molotkov, I. A., Wave propagation in an inhomogeneous half-space with the refractive index dependent on two coordinates, in: Wave Diffraction and Propagation Problems, Vol. 6, Leningrad Univ. (LGU) (1966).
2. Babich, V. M., A formal technique for constructing a short-wave asymptotic representation of the Green function, Trudy Matem. Inst. Steklov., Vol. 115 (in press).

HIGH-FREQUENCY ASYMPTOTIC BEHAVIOR OF THE SOLUTIONS OF THE WAVE EQUATION IN A PLANE WAVEGUIDE FORMED BY TWO CAUSTICS

Z. A. Yanson

The aim of the present article is to formulate an asymptotic representation for the solution of the equation

$$\left(\Delta + \frac{\omega^2}{c^2(x,z)}\right) u = 0, \tag{1}$$

in which

$$\Delta = \frac{\partial^2}{\partial x^2} + \frac{\partial^2}{\partial z^2}$$

[ω is the frequency, $\omega \to \infty$, and $c(x,z)$ is the wave propagation velocity], in the case in which the ray field has two envelopes, both caustics. The same physical characteristics, namely the presence of two envelopes for the ray system, are invested in the wave field u [solution of Eq. (1)] if the velocity depends only on one coordinate z and has a minimum at $z = z_0$. These conditions create a vertically inhomogeneous waveguide or channel in an acoustic medium. The asymptotic behavior of the wave equation solutions concentrated in the vicinity of the axis $z = z_0$ of such a waveguide has been treated in [1], in which the indicated solutions are formulated as a subsequence of functions, each of which is represented by an asymptotic series in reciprocal powers of the frequency ω.

Proceeding by analogy with [1], we shall seek those solutions of Eq. (1) which oscillate in a waveguide formed by caustics and decay exponentially outside it. Interest has been stimulated in the investigation of these particular solutions of Eq. (1) by a number of practical problems relating to underwater sound in the ocean and seismology and involving wave propagation via internal channels (waveguides) and the need to take proper account of the inhomogeneity of the medium along the wave path.

We shall formulate the solutions of Eq. (1) as formal asymptotic series in reciprocal powers of ω, where, as in [1], these solutions form a subsequence of functions

$$u_1, u_2, \ldots, u_m, \ldots,$$

for which the sign of m is related to the number of zeros of the function u_m in the waveguide cross section.

An asymptotic representation of the solutions of (1) concentrated near the axis of a plane waveguide has been formulated in [2], in which it was assumed that the width ν of the waveguide was small ($\nu \sim \omega^{-\frac{1}{2}}$). We shall be concerned in the present article with the formulation of only the principal (zero-order) term of the asymptotic series in reciprocal powers of the frequency ω, but we shall assume that the position of the observation point and the width between the caustics are arbitrary.

The equations derived below are uniform asymptotic equations and close to or far from the caustics can be compared with the well-known asymptotic representations of [3, 4, 5] for the solutions of Eq. (1).

The author is deeply indebted to V. M. Babich and V. S. Buldyrev for the statement of the problem and a discussion of the results.

§ 1. Statement of the Problem; Wave Field Far from the Caustics; Eikonal Equations

Let the following field of rays be given:

$$\vec{X} = \vec{X}(\alpha, \tau), \quad \vec{X} = (X, Z). \tag{1.1}$$

The coordinates α, τ are ray coordinates [3]; fixing the parameter α, we obtain a specific ray; fixing τ, we obtain a wave front. For the parameter τ, which characterizes a point on the ray, we adopt the transmit time of a perturbation along the ray:

$$\tau = \int_{\mathcal{M}_0}^{\mathcal{M}} \frac{ds}{c(x, z)}, \tag{1.2}$$

where $\mathcal{M}_0$ is a fixed point and $\mathcal{M}$ is a variable point on the ray. The function τ is the basic field function of the extremals, i.e., rays, and satisfies the eikonal equation

$$(\operatorname{grad} \tau)^2 = \frac{1}{c^2(x, z)}. \tag{1.3}$$

In the case of a vertically inhomogeneous medium, as shown in [1], the width of a waveguide formed by caustics depends on the frequency ω and the number of zeros of the function U_m in the waveguide cross section. It is reasonable in the case of a plane waveguide to expect the same wave pattern as in [1]; every ray tangent to one caustic must necessarily be tangent to the other caustic. The given problem is intrinsically related to the eigenfunction problem, and by analogy with [1] the width of the waveguide must be a function of both the number of zeros of U_m in the waveguide cross section and the frequenct ω.

We specify the equations for the two caustics generating the waveguide in the implicit form

$$\psi_m^{(\pm)}(x, z; \omega) = 0. \tag{1.4}$$

Two families of rays traveling in opposite directions relative to the caustics (1.4) are illustrated in Fig. 1 by solid and dashed curves, respectively. The form of the function $\psi_m^{(\pm)}$ will be established later.

We shall assume that the velocity $c(x,z)$ and the function $\vec{X}(\alpha,\tau)$ are sufficiently smooth functions of their arguments and that the tangency of the rays and caustic is a first-order tangency. Given the ray field (1.1), we assume the value of the eikonal τ on the caustic [in our problem either of the caustics (1.4)] as the parameter α characterizing a particular ray, so that the equation for the corresponding caustic acquires the form

$$\vec{X}_\alpha(\alpha, \tau)\Big|_{\tau=\alpha} = 0. \tag{1.5}$$

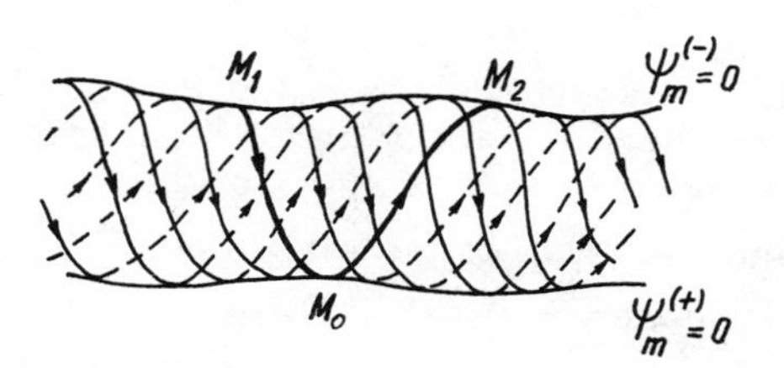

Fig. 1

1. Let the observation point $\mathcal{M}$ be located between the caustics (1.4), i.e., inside the waveguide, so that the equations of geometric optics are valid for the wave field U at that point. We deduce these equations at once by inspection of the already-established asymptotic equations for the solution of (1) that are uniformly applicable in the vicinity

of one caustic (see [4, 5, 6]). We adopt as our starting relations the representation obtained in [6], Chap. 2, for the wave field U in the vicinity of a caustic.

We have the following asymptotic equation for the solution of Eq. (1) in the vicinity of each of the caustics (1.4):

$$U^{(\pm)} = \left\{ v(-\omega^{2/3} m_\pm) \sum_{k=0}^{\infty} \frac{A_k^{(\pm)}(x,z)}{(-i\omega)^k} + i v'(-\omega^{2/3} m_\pm)\,\omega^{-1/3} \sum_{k=0}^{\infty} \frac{B_k^{(\pm)}(x,z)}{(-i\omega)^k} \right\} e^{i\omega l_0(x,z)}, \tag{1.6}$$

in which $v(t)$ is an Airy function and $m_\pm(x,z)$, $l_0(x,z)$ are functions having power series form in the vicinity of the caustics:

$$l_0(x,z) = \tau_0 + c^{-1}(M_0)\,x + \dots,$$

$$m_\pm(x,z) = \frac{2}{(\vec{X}_{\alpha\alpha})_z}\left(\frac{1}{2}(\vec{X}_{\alpha\alpha\alpha})_x\right)^{2/3}\Big|_{\alpha=\alpha_\pm} z + \dots. \tag{1.7}$$

The series expansions in Eq. (1.7) are made in a local coordinate system with center at the point M_0 of tangency of the given ray with the caustic. The quantity $\tau = \tau_0$ is the value of the eikonal on the caustic itself; the parameters $\alpha_\pm$ characterize the rays on the corresponding caustics.

Equations (1.7) determine the eikonal τ as a two-valued function of the point M. Two rays pass through the point M; these rays are represented by M_1M_4 and M_3M_2 in Fig. 2. If the ray M_1M_4 is regarded as a ray approaching, and M_3M_2 as a ray departing from the caustic $\psi_m^{(+)}(x,z,\omega) = 0$, we have the following equation for the eikonal:

$$\tau_+^{(1,2)} = l_0 \mp \frac{2}{3} m_+^{3/2}. \tag{1.8}$$

Regarding the ray M_1M_4, in turn, as a ray tangent to the caustic $\psi_m^{(-)}(x,z;\omega) = 0$, and M_3M_2 as a ray approaching the same caustic, we obtain

$$\tau_-^{(1,2)} = l_0 \mp \frac{2}{3} m_-^{3/2}. \tag{1.9}$$

The function m_- vanishes on the caustic $\psi_m^{(-)} = 0$, while the function m_+ vanishes on the caustic $\psi_m^{(+)} = 0$.

Expressions have been derived in [6] for the coefficients of the asymptotic series (1.6) for any order numbers k. For $k=0$ these coefficients give as the zeroth approximation for the solution of Eq. (1)

$$\left.\begin{aligned} A_0^{(\pm)} &= \frac{1}{2} m_\pm^{1/4} \sqrt{c(M)} \left(\frac{\chi_0^{(\pm)}(\alpha'_\pm)}{\sqrt{J'_\pm}} + \frac{\chi_0^{(\pm)}(\alpha''_\pm)}{\sqrt{J''_\pm}} \right), \\ B_0^{(\pm)} &= \frac{1}{2} m_\pm^{-1/4} \sqrt{c(M)} \left(\frac{\chi_0^{(\pm)}(\alpha''_\pm)}{\sqrt{J''_\pm}} - \frac{\chi_0^{(\pm)}(\alpha'_\pm)}{\sqrt{J'_\pm}} \right), \end{aligned}\right\}$$

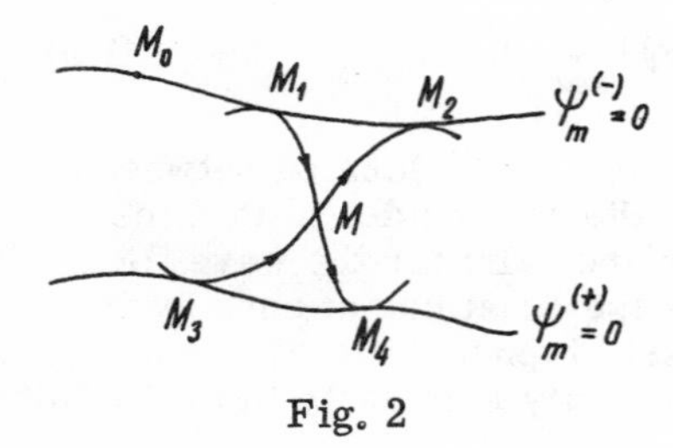

Fig. 2

where $J = |\vec{X}_\alpha(\alpha,\tau)|$ is the geometric divergence of the ray field and χ_0 is a function constant on the ray (in the form of a ray segment contained between the caustics in our problem). The single and double primes attached to the variables $\alpha^{\prime,\prime\prime}$ and $J^{\prime,\prime\prime}$ refer to the two ray families, respectively, i.e., those still tangent and those already departing from a particular one of the caustics (1.4).

2. Before we embark upon the description of the wave field far from the caustics (1.4) we need to clarify the dependence that must exist between the eikonals determined by Eqs. (1.8) and (1.9) if the caustics are to form a waveguide. In the given waveguide, as in [1], two normal ray congruences occur, one of which is represented in Fig. 1 by solid curves, the other by dashed curves. Fusing the two ray congruences along the caustics, we obtain a dualistic space, or "tube." The requirement of uniqueness on the part of $|U|$, where U is the wave field, leads to the following quantization on our dualistic space [6, 7]:

$$\omega \int_{\Gamma} \operatorname{grad} \tau \, d\vec{s} = 2\pi \left[m + \frac{\pi}{4} (\ell_c' - \ell_c'') \right], \quad m = 0, 1, 2, \ldots, \tag{1.10}$$

where Γ is a closed contour (nonhomotonic with zero) on the indicated space and $\ell_c = (\ell_c' - \ell_c'')$ is the caustic index, which is equal to the difference between the number of transitions of Γ through the caustics concurrently and countercurrently with the rays.

We place in the role of Γ the contour $\mathcal{M}\mathcal{M}_2\mathcal{M}_1\mathcal{M}_4\mathcal{M}_3\mathcal{M}$ (Fig. 2). We determine the eikonals on the rays $\mathcal{M}_1\mathcal{M}$ and $\mathcal{M}\mathcal{M}_2$ by Eqs. (1.8), and those on the rays $\mathcal{M}_3\mathcal{M}$ and $\mathcal{M}\mathcal{M}_4$ by Eqs. (1.9). The integral on the left-hand side of (1.10), i.e., the total eikonal increment, is equal to the following on our chosen contour:

$$\int_{\Gamma} \operatorname{grad} \tau \, d\vec{s} = 2 \left[\frac{2}{3} m_+^{3/2} (\mathcal{M}) + \frac{2}{3} m_-^{3/2} (\mathcal{M}) \right],$$

and the caustic index has a value $\ell_c = 2$. Thus, condition (1.10) has the effect that the "transverse" parts $m_\pm$ of the eikonals $\tau_\pm$ satisfy the relation

$$\frac{2}{3} \left[m_+^{3/2} (\mathcal{M}) + m_-^{3/2} (\mathcal{M}) \right] = \pi \sigma,$$

in which $\sigma = {m + 1/2}/\omega$ so that the eikonals $\tau_-^{(1,2)}$ and $\tau_+^{(1,2)}$ themselves are interrelated by the equation

$$\tau_-^{(1,2)} = \tau_+^{(2,1)} \mp \pi \sigma, \tag{1.11}$$

which also occurs in [1].

If the point $\mathcal{M}$ is on one of the caustics, say $\psi_m^{(+)} = 0$, and coincides with the point $\mathcal{M}_0$ in Fig. 1, while the contour Γ coincides with the contour $\mathcal{M}_1\mathcal{M}_0\mathcal{M}_2\mathcal{M}_1$, the quantization condition (1.10) assumes the form

$$\omega \Delta \tau_\pm = (m + \tfrac{1}{2}) \, 2\pi, \tag{1.12}$$

where the plus or minus sign depends on whether Eqs. (1.8) or (1.9) determine the values of the eikonals on the rays $\mathcal{M}_1\mathcal{M}_0$ and $\mathcal{M}_0\mathcal{M}_2$. Relation (1.12) means that the transit time difference between a perturbation via the ray $\mathcal{M}_1\mathcal{M}_0\mathcal{M}_2$ and via the caustic segment $\mathcal{M}_1\mathcal{M}_2$, in Fig. 1 must be equal to $(m + \frac{1}{2})T$, where $T = \frac{2\pi}{\omega}$ is the wave period. Different values of m correspond to waveguides of different widths.

In concluding the present section we consider the problem of how to compute the functions $\ell_0(x, z)$ and $m_\pm(x, z)$ far from the caustics (inside the waveguide), where Eqs. (1.7) can become inapplicable. If the eikonal τ is determined from Eq. (1.9), we have the following at a distance from the caustic $\psi_m^{(-)} = 0$:

$$\ell_0 = \frac{\tau_-^{(1)} + \tau_-^{(2)}}{2}, \quad \frac{2}{3} m_-^{3/2} = \frac{\tau_-^{(2)} - \tau_-^{(1)}}{2}, \tag{1.13}$$

where

$$\tau_-^{(1)} = \left(\int_{\mathcal{M}_0}^{\mathcal{M}_2} - \int_{\mathcal{M}}^{\mathcal{M}_2} \right) \frac{ds}{c},$$

$$\tau_-^{(2)} = \left(\int_{\mathcal{M}_0}^{\mathcal{M}_1} + \int_{\mathcal{M}_1}^{\mathcal{M}} \right) \frac{ds}{c}, \tag{1.13a}$$

and the point $\mathcal{M} = \mathcal{M}_0$ is an arbitrary fixed point on the caustic $\Psi_m^{(-)} = 0$ in Fig. 2.

3. Let the observation point $\mathcal{M}$ be located between the caustics, with $m_\pm > 0$. Replacing the Airy function in (1.6) by its asymptotic representation as $\omega \to \infty$, we obtain the following expression for $U^{(+)}(\mathcal{M})$:

$$U^{(+)}(\mathcal{M}) = \omega^{-\frac{1}{6}} e^{i\frac{\pi}{4}} \left\{ \chi_0^{(+)}(\alpha_+') \sqrt{\frac{c}{J_+'}}\, e^{i\omega\tau_+^{(1)}} + \chi_0^{(+)}(\alpha_+'') \sqrt{\frac{c}{J_+''}}\, e^{i\omega\tau_+^{(2)} - i\frac{\pi}{2}} \right\} \left\{ 1 + O\left(\frac{1}{\omega}\right) \right\}. \tag{1.14}$$

We interpret $U^{(-)}(\mathcal{M})$ as the expression obtained from (1.6) when the Airy function is replaced by its asymptotic representation; this expression differs from the original by the factor $e^{im\pi}$, i.e.,

$$U^{(-)}(\mathcal{M}) = \omega^{-\frac{1}{6}} e^{i(m+\frac{1}{4})\pi} \left\{ \chi_0^{(-)}(\alpha_-'') \sqrt{\frac{c}{J_-''}}\, e^{i\omega\tau_-^{(2)} - i\frac{\pi}{2}} + \chi_0^{(-)}(\alpha_-') \sqrt{\frac{c}{J_-'}}\, e^{i\omega\tau_-^{(1)}} \right\} \left\{ 1 + O\left(\frac{1}{\omega}\right) \right\}. \tag{1.15}$$

The principal terms of Eqs. (1.14) and (1.15) represent the sum of two components having a "ray" character. The functions $\chi_0 \sqrt{\frac{c}{J}}$ comprising the factors in front of the exponentials are interpreted as intensities on the rays whose eikonals are contained in the phases of the exponentials.

We now write the conditions for coincidence of the zeroth approximations given for the solution of Eq. (1) by the principal terms of Eqs. (1.14) and (1.15). With regard for relations (1.11) these conditions assume the form

$$\left. \begin{aligned} \frac{\chi_0^{(-)}(\alpha_-')}{\chi_0^{(+)}(\alpha_+'')} &= \sqrt{\frac{J_-'}{J_+''}}, \\ \frac{\chi_0^{(-)}(\alpha_-'')}{\chi_0^{(+)}(\alpha_+')} &= \sqrt{\frac{J_-''}{J_+'}}, \end{aligned} \right\} \tag{1.16}$$

in which $\alpha_-^{','''}$ and $\alpha_+^{','''}$ are the values of the eikonals τ_- and τ_+ at points of tangency with the caustics of the two rays passing through the point $\mathcal{M}$ in Fig. 2, so that $\tau_-^{(1)}(\mathcal{M}_2) = \alpha_-'$, $\tau_-^{(2)}(\mathcal{M}_1) = \alpha_-''$, $\tau_+^{(1)}(\mathcal{M}_4) = \alpha_+'$, $\tau_+^{(2)}(\mathcal{M}_3) = \alpha_+''$. The right-hand sides of relations (1.16) are known functions depending on the value of the eikonal at a variable point of the ray. The left-hand sides of the indicated equations depend only on the values of the eikonals τ_+ and τ_- at the points of tangency of one and the same ray ($\mathcal{M}_1 \mathcal{M}_4$ or $\mathcal{M}_3 \mathcal{M}_2$ in Fig. 2) with the caustics (1.4). In the next section we shall show that the ratio $\sqrt{\frac{J_-}{J_+}}$ on one and the same ray is in fact a function of the values of the eikonals or the function $\ell_0(x, z)$ at the points of tangency of the ray with the caustics.

Equations (1.16) therefore make it possible to fuse the principal terms of solutions of the form (1.6) inside the waveguide formed by the caustics, where the geometric-optical approximation is valid.

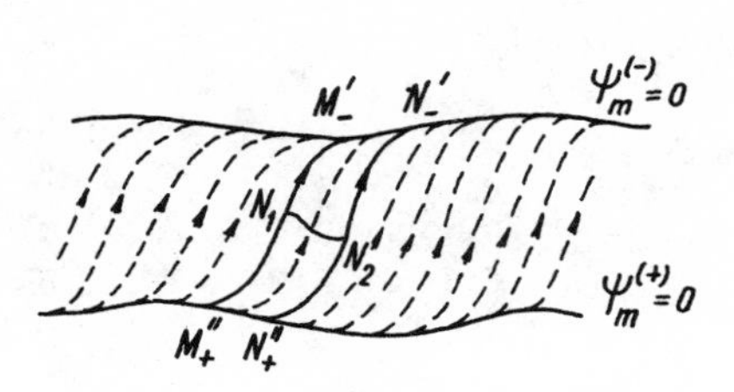

Fig. 3

§ 2. Derivation of Functional Equations for the Functions $\chi_0^{(+)}$ and $\chi_0^{(-)}$

Consider the normal congruence of rays (Fig. 3)

$$\vec{X} = \vec{X}(\alpha_-', \tau_-^{(1)}), \quad \vec{X} = (x, z). \tag{2.1}$$

The prime attached to the parameter α_-, which coincides with the value of $\tau_-^{(1)}$ on the caustic $\psi_m^{(-)} = 0$ signifies that we are dealing with rays traveling toward the indicated caustic. On the other hand, the same normal ray congruence has the form

$$\vec{X} = \vec{X}(\alpha_+'', \tau_+^{(2)}),^* \tag{2.2}$$

where the double prime on the parameter α_+ signifies that the congruence (2.2) consists of rays departing from the caustic $\psi_m^{(+)} = 0$; the parameter α_+ coincides with the value of $\tau_+^{(2)}$ on the same caustic.

On the congruence (2.1) [or, equivalently, on (2.2)] we investigate the ray tube $\mathcal{M}_-' \mathcal{M}_+'' N_+'' N_-'$ and let $N_1 N_2$ - be a certain position of the wave front on this tube. We see at once that

$$|N_1 N_2| = |\vec{X}_{\alpha_-'}(\alpha_-', \tau_-^{(1)})| \, \Delta\alpha_-' + O((\Delta\alpha_-')^2), \tag{2.3}$$

whereas

$$|N_1 N_2| = |\vec{X}_{\alpha_+''}(\alpha_+'', \tau_-^{(2)})| \, \Delta\alpha_+'' + O((\Delta\alpha_+'')^2), \tag{2.4}$$

in which

$$\alpha_-' = \tau_-^{(1)}(\mathcal{M}_-'), \quad \alpha_+'' = \tau_+^{(2)}(\mathcal{M}_+''),$$

$$\Delta\alpha_-' = (\tau_-^{(1)}(N_-') - \tau_-^{(1)}(\mathcal{M}_-')) > 0,$$

$$\Delta\alpha_+'' = (\tau_+^{(2)}(N_+'') - \tau_+^{(2)}(\mathcal{M}_+'')) > 0.$$

Equating (2.3) and (2.4) and letting $\Delta\alpha_-'$ tend to zero, we obtain

$$\frac{|\vec{X}_{\alpha_-'}|}{|\vec{X}_{\alpha_+''}|} = \frac{d\alpha_+''}{d\alpha_-'}, \tag{2.5}$$

where $\frac{d\alpha_+''}{d\alpha_-'}$ is the derivative of the function $\alpha_+'' = \alpha_+''(\alpha_-')$ with respect to α_-'.

Investigating in analogous fashion the other normal congruence

$$\vec{X} = \vec{X}(\alpha_-'', \tau_-^{(2)}) \tag{2.6}$$

of rays traveling from the caustic $\psi_m^{(-)} = 0$ to the caustic $\psi_m^{(+)} = 0$, we arrive at the relation

$$\frac{|\vec{X}_{\alpha_-''}|}{|\vec{X}_{\alpha_+'}|} = \frac{d\alpha_+'}{d\alpha_-''}. \tag{2.7}$$

---

*The functional dependence $\vec{X} = \vec{X}(\alpha, \tau)$ differs in Eqs. (2.1) and (2.2) in general, but so as not to complicate the equations we shall not assign different symbols to these functions.

Bearing in mind the equations

$$\jmath_- = |\vec{X}_{\alpha_-}(\alpha_-, \tau_-)|, \tag{2.8a}$$

$$\jmath_+ = |\vec{X}_{\alpha_+}(\alpha_+, \tau_+)|, \tag{2.8b}$$

we rewrite relations (2.5) and (2.7) in the form

$$\left.\begin{aligned} \frac{\jmath_-'}{\jmath_+''} &= \frac{d\alpha_+''}{d\alpha_-'}, \\ \frac{\jmath_-''}{\jmath_+'} &= \frac{d\alpha_+'}{d\alpha_-''}. \end{aligned}\right\} \tag{2.9}$$

It follows from Eqs. (2.9) that the right-hand sides of the above equations (1.16) do in fact depend on the values of τ_- and τ_+ at the moving point of the ray and are a function of α_- and α_+.

The derivatives $\frac{d\alpha_+''}{d\alpha_-'}$ and $\frac{d\alpha_+'}{d\alpha_-''}$ can be determined as follows. The total increment of the eikonal, $\Delta\tau_-^{(1)}$ (or $\Delta\tau_+^{(2)}$), is equal to zero on the closed path $\mathcal{M}_-'\mathcal{M}_+''N_+''N_-'\mathcal{M}'$, i.e.,

$$-\int_{\mathcal{M}_+''}^{\mathcal{M}_-'} \frac{ds}{c} + \Delta\alpha_+'' + \int_{N_+''}^{N_-'} \frac{ds}{c} - \Delta\alpha_-' = 0.$$

Hence, letting $\Delta\alpha_-'$ tend to zero, we find

$$\frac{d\alpha_+''}{d\alpha_-'} = 1 - \frac{d}{d\alpha_-'} \int_{\mathcal{M}_+''}^{\mathcal{M}_-'} \frac{ds}{c}, \tag{2.10}$$

in which we have the notation

$$\frac{d}{d\alpha_-'} \int_{\mathcal{M}_+''}^{\mathcal{M}_-'} \frac{ds}{c} = \lim_{\Delta\alpha_-' \to 0} \frac{\int_{N_+''}^{N_-'} \frac{ds}{c} - \int_{\mathcal{M}_+''}^{\mathcal{M}_-'} \frac{ds}{c}}{\Delta\alpha_-'}.$$

For the other ray field (2.6) we obtain analogously

$$\frac{d\alpha_+'}{d\alpha_-''} = 1 + \frac{d}{d\alpha_-''} \int_{\mathcal{M}_-''}^{\mathcal{M}_+'} \frac{ds}{c}. \tag{2.11}$$

Equations (1.16) were obtained in the region between the caustics (1.4) in which the geometric-optical approximation is valid and express the relationship between the values of the functions $\chi_0^{(+)}$ and $\chi_0^{(-)}$ on one and the same ray contained between the caustics, under the condition that the quantization condition (1.12) is fulfilled in the waveguide. We assume as a hypothesis that relations (1.16) are valid everywhere inside the waveguide up to the caustics themselves. In this case we reveal the true meaning of relations (1.16).

We pick a point $\mathcal{M}$ on the ray $\mathcal{M}_1\mathcal{M}_0$ in Fig. 1. We divide the two equations (1.16) term by term and find the limit of the ratio as $\mathcal{M} \to \mathcal{M}_0$ along the ray $\mathcal{M}_1\mathcal{M}_0$. Taking Eqs. (2.9), (2.10), and (2.11) into account and assuming in the limit that $\alpha_+' = \alpha_+'' = \alpha_+$, where α_+ is the value of the eikonal τ_+ (more precisely, of the function ℓ_0) at the point $\mathcal{M}_0$, we deduce the relation

$$\chi_0^{(-)}(\alpha_-')\Big|_{\mathcal{M}=\mathcal{M}_2} = \chi_0^{(-)}(\alpha_-'')\Big|_{\mathcal{M}=\mathcal{M}_1} \times \sqrt{\frac{1-\frac{d}{d\alpha_-'}\int_{\mathcal{M}_0}^{\mathcal{M}_2}\frac{ds}{c}}{1+\frac{d}{d\alpha_-''}\int_{\mathcal{M}_1}^{\mathcal{M}_0}\frac{ds}{c}}}. \quad (2.12)$$

Equation (2.12) represents a functional equation of the form

$$\chi_0^{(-)}(\alpha_- + L(\alpha_-)) = \chi_0^{(-)}(\alpha_-)\, f(\alpha_-), \quad (2.13)$$

in which α_- is the value of the eikonal τ_-, or the function $\ell_0(x, z)$, on the caustic $\psi_m^{(-)} = 0$; $f(\alpha_-)$ is the radical on the right-hand side of Eq. (2.12), and $L(\alpha_-)$ is a function equal to the transit time of a perturbation along the caustic $\psi_m^{(-)} = 0$ between the two nearest tangency points of the same ray with the indicated caustic (these points are $\mathcal{M}_1$ and $\mathcal{M}_2$ in Fig. 1). Clearly, an equation of the form (2.13) can also be deduced for the function $\chi_0^{(+)}$ if the point $\mathcal{M}$ in relations (1.16) somehow moves toward a point on the caustic $\psi_m^{(-)} = 0$. We then obtain

$$\chi_0^{(+)}(\alpha_+ + L(\alpha_+)) = \chi_0^{(+)}(\alpha_+)\, f(\alpha_+), \quad (2.14)$$

where α_+ is the value of the eikonal τ_+ on the caustic $\psi_m^{(+)} = 0$.

We note that a functional equation of the type (2.13) [or, accordingly, (2.14)] relates the values of the function $\chi_0^{(-)}$ (or $\chi_0^{(+)}$) at any two points (M_1 and M_N) of tangency of the ray with the caustic $\psi_m^{(-)} = 0$ (or $\psi_m^{(+)} = 0$). The function $f(\alpha_-)$ [or, analogously, $f(\alpha_+)$], which accounts for the influence of the second caustic on the value of $\chi_0^{(-)}$ (or $\chi_0^{(+)}$), turns out in this case to be the product of radicals

$$\prod_{i=1}^{N} \sqrt{\frac{1-\frac{d}{d\alpha_-'}\int_{M_0^{(i)}}^{M_{i+1}} ds/c}{1+\frac{d}{d\alpha_-''}\int_{M_i}^{M_0^{(i)}} ds/c}},$$

in which M_i and M_{i+1}, $i = 1, 2, \ldots, N$ are tangency points of the ray with the caustic $\psi_m^{(-)} = 0$, while $M_0^{(i)}$, $i = 1, 2, \ldots, N$ are tangency points of the same ray with the second caustic $\psi_m^{(+)} = 0$.

We now consider two aspects of Eq. (2.13).

1) We ascertain to what Eq. (2.13) reduces in the case of an "equidistant" waveguide (we shall define this type of waveguide presently).

2) We consider how the function, say $\chi_0^{(-)}$, can be specified on an arbitrary segment $\Delta\ell$ of the caustic $\psi_m^{(-)} = 0$ so that $\Delta\ell < L(\alpha_-)$. It follows from Eq. (2.13) that the function $\chi_0^{(-)}$ will then be known on all segments of the indicated caustic that are tangent with rays defined by the given distribution of the function $\chi_0^{(-)}$.

If the transit time along rays of either of the two congruences forming the waveguide from one caustic to the other is equal to a constant, we call the waveguide an "equidistant" waveguide. Such waveguides are characterized in Eqs. (2.13) and (2.14) by $L(\alpha_\mp) = L = \text{const}$ and $f(\alpha_\mp) = 1$, so that the indicated equations acquire the form

$$\chi_0^{(\mp)}(\alpha_\mp + L) = \chi_0^{(\mp)}(\alpha_\mp),$$

i.e., the functions $\chi_0^{(\mp)}$ must be periodic functions with period L. A special case of "equidistant" waveguide is the vertically inhomogeneous waveguide treated in [1]. This type of waveguide is characterized by $\chi_0^{(-)} = \chi_0^{(+)} = \text{const}$. We defer the verification of the latter assertion, plus the solution of the problem of how to specify the function $\chi_0^{(-)}$ (or $\chi_0^{(+)}$) on any segment of the caustic $\Psi_m^{(-)} = 0$ (or $\Psi_m^{(+)} = 0$) to the ensuing sections, in which we construct uniform asymptotic equations for the solution of Eq. (1).

§ 3. Formation of Uniform Asymptotic Equations in a Plane Waveguide

1. For our standard problem we use the problem of wave propagation in a vertically inhomogeneous waveguide of the type investigated in [1]. We seek the solution of Eq. (1) in a form similar to the solution of the standard problem [1], viz.,

$$U_m(x,z) \sim \Big\{ \mathcal{D}_m(\omega^{\frac{1}{2}}\varphi(x,z)) \Big[A_0(x,z) + \frac{A_1(x,z)}{(-i\omega)} + \frac{A_2(x,z)}{(-i\omega)^2} + \ldots\Big] +$$

$$+ \beta\omega^{-\alpha}\mathcal{D}_m'(\omega^{\frac{1}{2}}\varphi(x,z))\Big[B_0(x,z) + \frac{B_1(x,z)}{(-i\omega)} + \frac{B_2(x,z)}{(-i\omega)^2} + \ldots\Big]\Big\} e^{i\omega \ell_0(x,z)}, \tag{3.1}$$

$$m = 0, 1, 2, \ldots,$$

where $\mathcal{D}_m(\zeta)$ is a parabolic-cylindrical function satisfying the equation

$$\mathcal{D}_m''(\omega^{\frac{1}{2}}\varphi) + \omega\left(\frac{m+\frac{1}{2}}{\omega} - \frac{\varphi^2}{4}\right)\mathcal{D}_m(\omega^{\frac{1}{2}}\varphi) = 0, \tag{3.2}$$

φ, ℓ_0, A_k, and B_k are unknown functions, and α and β are constants. The prime attached to the function $\mathcal{D}_m'$ in Eq. (3.1) denotes differentiation with respect to the argument $\zeta = \omega^{\frac{1}{2}}\varphi$.

The substitution of expression (3.1) into Eq. (1) yields

$$\mathcal{F}_1(x,z;\omega)\mathcal{D}_m(\omega^{\frac{1}{2}}\varphi) + \mathcal{F}_2(x,z;\omega)\mathcal{D}_m'(\omega^{\frac{1}{2}}\varphi) = 0, \tag{3.3}$$

where the functions $\mathcal{F}_1$ and $\mathcal{F}_2$ have the form

$$\mathcal{F}_1(x,z;\omega) = (-i\omega)^2\Big\{\Big[\Big((\nabla\varphi)^2\Big(\sigma - \frac{\varphi^2}{4}\Big) + (\nabla\ell_0)^2 - \frac{1}{c^2(x,z)}\Big)\Big(A_0 + \frac{A_1}{(-i\omega)} + \ldots\Big) +$$

$$+ 2\beta i\Big(\sigma - \frac{\varphi^2}{4}\Big)(\nabla\ell_0, \nabla\varphi)\Big(B_0 + \frac{B_1}{(-i\omega)} + \ldots\Big)\Big] - \frac{1}{(-i\omega)}\Big[i\beta\nabla\Big(\Big(\sigma - \frac{\varphi^2}{4}\Big)\nabla\varphi\Big) \times$$

$$\times\Big(B_0 + \frac{B_1}{(-i\omega)} + \ldots\Big) + 2\beta i\Big(\sigma - \frac{\varphi^2}{4}\Big)\nabla\varphi\Big(\nabla B_0 + \frac{\nabla B_1}{(-i\omega)} + \ldots\Big) + 2\nabla\ell_0\Big(\nabla A_0 + \tag{3.4}$$

$$+ \frac{\nabla A_1}{(-i\omega)} + \ldots\Big) + \Delta\ell_0\Big(A_0 + \frac{A_1}{(-i\omega)} + \ldots\Big)\Big] + \frac{1}{(-i\omega)^2}\Big(\Delta A_0 + \frac{\Delta A_1}{(-i\omega)} + \ldots\Big)\Big\},$$

$$\mathcal{F}_2(x,z;\omega) = \beta(-i\omega)^{2-\alpha}\Big\{\Big[\Big((\nabla\varphi)^2\Big(\sigma - \frac{\varphi^2}{4}\Big) + (\nabla\ell_0)^2 - \frac{1}{c^2(x,z)}\Big)\Big(B_0 + \frac{B_1}{(-i\omega)} + \dots\Big) -$$
$$-2\,i/\beta\,\omega^{-\frac{1}{2}+\alpha}(\nabla\ell_0,\nabla\varphi)\Big(A_0 + \frac{A_1}{(-i\omega)} + \dots\Big)\Big] + \frac{1}{(-i\omega)}\Big[\omega^{-\frac{1}{2}+\alpha}\,i/\beta\;2\nabla\varphi\Big(\nabla A_0 + \tag{3.5}$$
$$+\frac{\nabla A_1}{(-i\omega)} + \dots\Big) - 2\nabla\ell_0\Big(\nabla B_0 + \frac{\nabla B_1}{(-i\omega)} + \dots\Big) + \omega^{-\frac{1}{2}+\alpha}\frac{i}{\beta}\Delta\varphi\Big(A_0 + \frac{A_1}{(-i\omega)} + \dots\Big) -$$
$$-\Delta\ell_0\Big(B_0 + \frac{B_1}{(-i\omega)} + \dots\Big)\Big] + \frac{1}{(-i\omega)^2}\Big(\Delta B_0 + \frac{\Delta B_1}{(-i\omega)} + \dots\Big)\Big\},$$

and the following notation is introduced:

$$\sigma = \frac{m + \frac{1}{2}}{\omega}.$$

The values of $\varphi_{1,2} = \mp 2\sqrt{\sigma}$ are turning points of the differential equation (3.2), in which ω is a large parameter. Upon transition across the lines

$$\varphi(x,z) \pm 2\sqrt{\sigma} = 0 \tag{3.6}$$

the nature of the asymptotic behavior of Eq. (3.2) changes. The function $\mathcal{D}_m(\omega^{\frac{1}{2}}\varphi)$ oscillates in the strip $|\varphi| < 2\sqrt{\sigma}$ and decays exponentially for $|\varphi| > 2\sqrt{\sigma}$. We identify the lines (3.6) with the caustics, i.e., with the envelopes of the family of rays and assume, in accordance with (1.4), that $\Psi_m^{(\pm)} = -\varphi(x,z) \pm 2\sqrt{\sigma}$.

By virtue of the linear independence of the functions $\mathcal{D}_m(\varsigma)$ and $\mathcal{D}_m'(\varsigma)$ it follows from (3.3) that the following relations must be satisfied:

$$\mathcal{F}_1(x,z;\omega) = 0, \tag{3.7a}$$

$$\mathcal{F}_2(x,z,\omega) = 0. \tag{3.7b}$$

In (3.4) and (3.5) we set the parameters α and β equal to $\alpha = \frac{1}{2}$ and $\beta = -i$ (see also Eq. (2) in [8]) and require that Eqs. (3.7a) and (3.7b) be approximately fulfilled with error $O(1)$ and $O(\omega^{-\frac{1}{2}})$. This will be true if the terms in powers $(-i\omega)^0$ and $(-i\omega)^{-1}$ in the braces in (3.4) and (3.5) are equal to zero. The extinction of the higher-power terms, i.e., in $(-i\omega)^0$ leads to the equations

$$\left.\begin{aligned}&(\nabla\ell_0)^2 + \Big(\sigma - \frac{\varphi^2}{4}\Big)(\nabla\varphi)^2 = \frac{1}{c^2(x,z)},\\ &(\nabla\ell_0,\nabla\varphi) = 0,\end{aligned}\right\} \tag{3.8}$$

while the extinction of terms in $(-i\omega)^{-1}$ yields the following system of equations for the coefficients A_0 and B_0:

$$\begin{aligned}&\mathcal{L}_2 A_0 + \mathcal{L}_3 B_0 = 0,\\ &\mathcal{L}_1 A_0 + \mathcal{L}_2 B_0 = 0,\end{aligned} \tag{3.9}$$

in which the operators $\mathcal{L}_j$, $j = 1, 2, 3$, have the form

$$\mathcal{L}_1 f = f\Delta\varphi + 2\nabla\varphi\nabla f,$$

$$\mathcal{L}_2 f = 2\nabla \ell_0 \nabla f + f\Delta \ell_0,$$

$$\mathcal{L}_3 f = \nabla\left(\left(\sigma - \frac{\varphi^2}{4}\right)\nabla\varphi\right) f + 2\left(\sigma - \frac{\varphi^2}{4}\right)\nabla\varphi\nabla f.$$

We note that the system of equations (3.9) is very similar to the corresponding system obtained for the coefficients A_0 and B_0 in [6], Chap. 2, for the case of a single caustic.

2. A system of equations analogous to (3.8) has been obtained in [8] and, as in [8], may be written in the form of a single equation, namely the eikonal equation (1.3), if we set

$$\tau = \ell_0(x, z) \mp \int \sqrt{\sigma - \frac{\varphi^2}{4}}\, d\varphi. \tag{3.10}$$

The integral in (3.10) can, in turn, be normalized so that the eikonal τ will be a single-valued function, either: a) on the caustic $\varphi = -2\sqrt{\sigma}$, or b) on the caustic $\varphi = 2\sqrt{\sigma}$.

In case a) we have

$$\tau \equiv \tau_-^{(1,2)} = \ell_0(x, z) \mp \int_{-2\sqrt{\sigma}}^{\varphi} \sqrt{\sigma - \frac{\varphi^2}{4}}\, d\varphi. \tag{3.11}$$

In case b) we have

$$\tau \equiv \tau_+^{(1,2)} = \ell_0(x, z) \mp \int_{\varphi}^{2\sqrt{\sigma}} \sqrt{\sigma - \frac{\varphi^2}{4}}\, d\varphi. \tag{3.12}$$

The functions $\ell_0(x, z)$ and $\varphi(x, z)$ are single-valued functions of the point M with coordinates (x, z) in Eqs. (3.11) and (3.12).

Equations (3.11) and (3.12) become exactly analogous to (1.8) and (1.9) if we put

$$\left.\begin{aligned} \frac{2}{3} m_-^{3/2} &\equiv \int_{-2\sqrt{\sigma}}^{\varphi} \sqrt{\sigma - \frac{\varphi^2}{4}}\, d\varphi, \\ \frac{2}{3} m_+^{3/2} &\equiv \int_{\varphi}^{2\sqrt{\sigma}} \sqrt{\sigma - \frac{\varphi^2}{4}}\, d\varphi. \end{aligned}\right\} \tag{3.13}$$

Relations (1.11) and (1.12), which were obtained in § 1, are automatically fulfilled. Thus, the quantity $\Delta\tau_\pm$ on the left-hand side of (1.12), i.e., the total increment of the eikonals $\tau_\pm$ on the contour $M_1 M_0 M_2 M_1$ in Fig. 1, turn out to be equal to

$$\Delta\tau_\pm = 2\int_{-2\sqrt{\sigma}}^{2\sqrt{\sigma}} \sqrt{\sigma - \frac{\varphi^2}{4}}\, d\varphi = 2\pi\sigma,$$

where $\sigma = m + \frac{1}{2}/\omega$.

By virtue of (3.8) the lines $\ell_0 = \text{const}$ and $\varphi = \text{const}$ form an orthonormal net on the plane (x, z), where $\tau = \ell_0(x, z)$ on the caustics $\varphi \pm 2\sqrt{\sigma} = 0$. The functions $\ell_0(x, z)$ and $\varphi(x, z)$ can be calculated near the caustics in the power series form (1.7), taking into account the notation of (3.13). Far from the caustics (inside the waveguide), on the other hand, where the series (1.7) can become divergent, the functions ℓ_0 and φ are conveniently calculated according to Eqs. (1.13) and (1.13a), provided expressions (3.11) are assumed for the determination of the eikonal.

3. We shall assume that the functions ℓ_0 and φ are known and that the values of the eikonals on the rays are determined in terms of these functions according to Eqs. (3.11). We next consider the system (3.9). We multiply the first equation of this system by $\left(\sigma - \frac{\varphi^2}{4}\right)^{-\frac{1}{4}}$ and multiply the second equation by $\left(\sigma - \frac{\varphi^2}{4}\right)^{\frac{1}{4}}$. Adding and subtracting the ensuing equations, we write the system (3.9) in the form

$$\left.\begin{aligned} &2\nabla\tau_-^{(1)}\nabla\Phi_-' + \Phi_-'\Delta\tau_-^{(1)} = 0, \\ &2\nabla\tau_-^{(2)}\nabla\Phi_-'' + \Phi_-''\Delta\tau_-^{(2)} = 0, \end{aligned}\right\} \tag{3.14}$$

in which the following notation is incorporated:

$$\Phi_-^{','' } = \left(\sigma - \frac{\varphi^2}{4}\right)^{-\frac{1}{4}} A_0 \mp \left(\sigma - \frac{\varphi^2}{4}\right)^{\frac{1}{4}} B_0. \tag{3.15}$$

Integration of the system (3.15) leads to the following zeroth-approximation ray-theoretic equations:

$$\left.\begin{aligned} &\Phi_-' = \chi_0^{(-)}(\alpha_-')\sqrt{\frac{c}{J_-'(\alpha_-', \tau_-^{(1)})}}, \\ &\Phi_-'' = \tilde{\chi}_0^{(-)}(\alpha_-'')\sqrt{\frac{c}{J_-''(\alpha_-'', \tau_-^{(2)})}}, \end{aligned}\right\} \tag{3.16}$$

in which Φ_-' is the perturbation intensity in the field of rays traveling toward the caustic $\varphi + 2\sqrt{\sigma} = 0$ and Φ_-'' is the intensity on rays departing from the same caustic. In Eqs. (3.16) the function $\chi_0^{(-)}$ (or $\tilde{\chi}_0^{(-)}$) is a certain arbitrary function that is constant on the ray segment contained between the caustics. We adopt the value of the eikonal τ_- at tangency points of the rays with the caustic $\varphi + 2\sqrt{\sigma} = 0$ as the parameter α_- characterizing the ray (in our case the ray segment), as in §1. The quantities $J_-^{','' }$ represent the geometric divergence of the corresponding ray field, where Eq. (2.8a) is valid.

For the coefficients A_0 and B_0 we have from (3.15)

$$\left.\begin{aligned} &A_0 = \left(\sigma - \frac{\varphi^2}{4}\right)^{1/4} \frac{\Phi_-' + \Phi_-''}{2}, \\ &B_0 = \left(\sigma - \frac{\varphi^2}{4}\right)^{-\frac{1}{4}} \frac{\Phi_-'' - \Phi_-'}{2}. \end{aligned}\right\} \tag{3.16a}$$

As in [6], in order to ensure boundedness of B_0 on the caustic $\varphi + 2\sqrt{\sigma} = 0$, where $J_- = 0$, we must require that

$$\chi_0^{(-)} \equiv \tilde{\chi}_0^{(-)}.$$

The coefficient A_0 is also bounded on the caustic $\varphi + 2\sqrt{\sigma} = 0$ on account of

$$\lim_{\tau_-^{(1,2)} \to \alpha_-^{','' }} \frac{\sigma - \frac{\varphi^2}{4}}{(J_-^{','' })^2} = \text{const.} < \infty, \quad \text{const} \neq 0.^* \tag{3.17}$$

*The existence of the limit of the ratio $\left(\sigma - \frac{\varphi^2}{4}\right)/J_\mp^2$ as the point $\mathcal{M}$ tends along the ray to the caustics $\psi_m^{(\mp)} = 0$ follows from the analysis of [6], Chap. 2.

We now return to the system (3.9). We perform the same operations on this system as brought it to the form (3.14), except that now we assume that the values of the eikonals on the rays are determined in terms of the functions ℓ_0 and φ according to Eq. (3.12). The system (3.9) is thereby converted to the form

$$\left.\begin{aligned} &2\nabla\tau_+^{(1)}\nabla\Phi_+' + \Phi_+'\Delta\tau_+^{(1)} = 0, \\ &2\nabla\tau_+^{(2)}\nabla\Phi_+'' + \Phi_+''\Delta\tau_+^{(2)} = 0, \end{aligned}\right\} \tag{3.18}$$

in which

$$\Phi_+^{','' } = \left(\sigma - \frac{\varphi^2}{4}\right)^{-\frac{1}{4}} A_0 \pm \left(\sigma - \frac{\varphi^2}{4}\right)^{\frac{1}{4}} B_0. \tag{3.19}$$

Integrating the system (3.18) in the ray coordinates (α_+, τ_+), we obtain

$$\left.\begin{aligned} &\Phi_+' = \chi_0^{(+)}(\alpha_+')\sqrt{\frac{c}{J_+'(\alpha_+', \tau_+^{(1)})}}, \\ &\Phi_+'' = \tilde{\chi}_0^{(+)}(\alpha_+'')\sqrt{\frac{c}{J_+''(\alpha_+'', \tau_+^{(2)})}}. \end{aligned}\right\} \tag{3.20}$$

In Eqs. (3.20) $J_+^{','' }$ is the respective geometric divergence of rays traveling toward and moving away from the caustic $\varphi - 2\sqrt{\sigma} = 0$ [Eq. (2.8b) holds for J_+]; the parameter α_+ again characterizes the ray segment between the caustics and coincides with the value of the eikonal τ_+ at the tangency point of the ray with the caustic $\varphi - 2\sqrt{\sigma} = 0$. On the caustic $\varphi - 2\sqrt{\sigma} = 0$ itself, the divergence J_+ is equal to zero for $\tau_+ = \alpha_+$, and the coefficients A_0 and B_0 turn out to be bounded under the condition

$$\chi_0^{(+)} = \tilde{\chi}_0^{(+)}.$$

Inasmuch as the coefficients A_0 and B_0 in Eqs. (3.15) and (3.19) must have the same values, the following equations must hold at any point between the caustics and on the caustics themselves:

$$\Phi_-' = \Phi_+'', \qquad \Phi_-'' = \Phi_+', \tag{3.20a}$$

which lead to the relations (1.16) already derived in § 1. The hypothesis assumed in § 2 with regard to the validity of relations (1.16) at any point between the caustics has now become a necessary condition by virtue of (3.20a). Therefore, the function $\chi_0^{(-)}$ in (3.16) (or, analogously, $\chi_0^{(+)}$) must be a solution of the functional equation (2.13) [or (2.14)].

We now show that if the function $\chi_0^{(-)}$ is a solution of Eq. (2.13) [implying that relations (1.16) are fulfilled at any point between the caustics], then the coefficients A_0 and B_0, determined according to Eqs. (3.16a) and (3.16) and bounded on the caustic $\Psi_m^{(-)} = 0$, remain bounded functions when the point $\mathcal{M}$ tends to the caustic $\Psi_m^{(+)} = 0$, say, along the ray $\mathcal{M}_1\mathcal{M}_4$ in Fig. 2.

We have

$$A_0 = \frac{1}{2}\left(\sigma - \frac{\varphi^2}{4}\right)^{\frac{1}{4}}\sqrt{\frac{c}{J_-''}}\,\chi_0(\alpha_-'')\left\{1 + \frac{\chi_0^{(-)}(\alpha_-')}{\chi_0^{(-)}(\alpha_-'')}\sqrt{\frac{J_-''}{J_-'}}\right\}.$$

On account of (1.16)

$$\lim_{M\to M_4,\ \alpha_+''\to\alpha_+'} A_0 = \frac{\sqrt{c}}{2}\chi_0^{(+)}(\alpha_+')\lim\frac{(\sigma-\frac{\varphi^2}{4})^{\frac{1}{4}}}{\sqrt{\mathcal{J}_+'}}\left\{1+\sqrt{\frac{\mathcal{J}_+'}{\mathcal{J}_+''}}\,\frac{\chi_0^{(+)}(\alpha_+'')}{\chi_0^{(+)}(\alpha_+')}\right\},$$

where

$$\lim_{\alpha_+''\to\alpha_+'}\frac{\mathcal{J}_+'}{\mathcal{J}_+''}=1$$

(see [6], Chap. 2).

The coefficient A_0 is thus bounded at points of the caustic $\varphi-2\sqrt{\sigma}=0$; this fact ensues from the existence of a finite limit for the expression $\sigma-\frac{\varphi^2}{4}/\mathcal{J}_+^{\prime 2}$; see (3.17). Analogously

$$\lim_{M\to M_4,\ \alpha_+''\to\alpha_+'} B_0 = \frac{\sqrt{c}}{2}\chi_0^{(+)}(\alpha_+')\lim\frac{(\sigma-\frac{\varphi^2}{4})^{\frac{1}{4}}}{\sqrt{\mathcal{J}_+'}}\left\{\frac{1-\sqrt{\frac{\mathcal{J}_+'}{\mathcal{J}_+''}}\,\frac{\chi_0^{(+)}(\alpha_+'')}{\chi_0^{(+)}(\alpha_+')}}{(\sigma-\frac{\varphi^2}{4})^{\frac{1}{2}}}\right\}, \tag{3.21}$$

i.e., the determination of the coefficient B_0 at points of the caustic $\varphi-2\sqrt{\sigma}=0$ requires that we resolve an indeterminary of the form $0/0$ for the expression in the braces in (3.21), and this yields a finite value for the coefficient B_0 at points of the indicated caustic.

The end result of the present section may be phrased in terms of the following theorem.

Theorem. If a given field of rays $\vec{X}=\vec{X}(\alpha_-,\tau_-)$ has two envelopes, i.e., forms a waveguide, then expressions of the form

$$U_m=\left\{D_m(\omega^{\frac{1}{2}}\varphi(x,z))\left[A_0(x,z)+O\left(\tfrac{1}{\omega}\right)\right]-i\omega^{-\frac{1}{2}}D_m'(\omega^{\frac{1}{2}}\varphi(x,z))\left[B_0(x,z)+O\left(\tfrac{1}{\omega}\right)\right]\right\}e^{(\omega\, l_0(x,z))}, \tag{3.22}$$

where D_m is a parabolic-cylindrical function and the functions φ and l_0 are expressed in terms of the values of the eikonal τ_- according to Eq. (3.11), will be a uniform asymptotic representation (zeroth approximation) of the solutions of Eq. (1) if and only if the function $\chi_0^{(-)}$ in terms of which A_0 and B_0 are expressed according to the equations

$$A_0=(\sigma-\tfrac{\varphi^2}{4})^{\frac{1}{4}}\frac{\sqrt{c}}{2}\left(\frac{\chi_0^{(-)}(\alpha_-')}{\sqrt{\mathcal{J}_-'}}+\frac{\chi_0^{(-)}(\alpha_-'')}{\sqrt{\mathcal{J}_-''}}\right),$$

$$B_0=\left(\sigma-\frac{\varphi^2}{4}\right)^{-\frac{1}{4}}\frac{\sqrt{c}}{2}\left(\frac{\chi_0^{(-)}(\alpha_-'')}{\sqrt{\mathcal{J}_-''}}-\frac{\chi_0^{(-)}(\alpha_-')}{\sqrt{\mathcal{J}_-'}}\right),$$

is a solution of the functional equation (2.13).

An analogous theorem may be formulated in terms of the function $\chi_0^{(+)}$ when the ray field is specified in the form $\vec{X}=\vec{X}(\alpha_+,\tau_+)$.

We conclude the present section with a verification of the fact that the following is indeed true in a vertically inhomogeneous waveguide:

$$\chi_0^{(-)}=\chi_0^{(+)}=\text{const.} \tag{3.23}$$

A comparison of Eq. (3.22) with (1.16) [1] brings out the following expression for the intensities on the rays in the case of a vertically inhomogenous medium:

$$\Phi_-^{I,II} = \Phi_+^{II,I} = \text{const} \times \frac{\sqrt{c(z)}}{\sqrt[4]{1-\nu_0 c^2(z)}}, \tag{3.24}$$

in which ν_0 is the separation constant for the separation of variables in Eq. (1). But the quantity $\sqrt{1-\nu_0 c^2(z)}$ is the divergence of the ray field in the case $c = c(z)$. This fact is readily established by integrating the familiar (see [6], Chap. 1) differential equation

$$\frac{d}{d\tau} \ln \frac{\gamma}{c} = c^2 \Delta \tau$$

for the divergence γ. But then (3.23) is a consequence of (3.24).

§ 4. Physical Interpretation of the Resulting Equations

Let the observation point $\mathcal{M}$ be situated between the caustics in the domain of oscillation of the solution (3.22), so that the following inequality holds:

$$\sigma \gg \frac{\varphi^2}{4}. \tag{4.1}$$

We have under condition (4.1)

$$\int_{-2\sqrt{\sigma}}^{\varphi} \sqrt{\sigma - \frac{\varphi^2}{4}}\, d\varphi = \sqrt{\sigma}\varphi + \frac{\pi\sigma}{2} + O\left[\sigma\left(\frac{\varphi}{2\sqrt{\sigma}}\right)^3\right]. \tag{4.2}$$

We replace the parabolic-cylindrical function $\mathcal{D}_m$ and its derivative $\mathcal{D}_m'$ by their asymptotic representations in the domain (4.1) [9] and, using (4.2), we obtain for the principal term of expression (3.22)

$$U_m(x,z) \approx \sqrt[4]{\sigma} \frac{\Gamma(m+1)}{2\Gamma\left(\frac{m}{2}+1\right)} e^{i\pi m + \frac{\pi}{4} i} \left\{ \left(\sigma^{-\frac{1}{4}} A_0 + \sigma^{\frac{1}{4}} B_0\right) e^{i\omega\tau_-^{(2)} - i\frac{\pi}{2}} + \left(\sigma^{-\frac{1}{4}} A_0 - \sigma^{\frac{1}{4}} B_0\right) e^{i\omega\tau_-^{(1)}} \right\}, \tag{4.3}$$

where $\Gamma(t)$ is the gamma function and the eikonals $\tau_-^{(1,2)}$ are determined by Eqs. (3.11). Expression (4.3) is the analog of Eq. (1.15), and the quantity contained in the braces is the sum of a wave departing from the caustic $\varphi + 2\sqrt{\sigma} = 0$ and a wave approaching it. The amplitudes of these waves are the approximate values of the intensities $\Phi_-^{I,II}$ in (3.15) under condition (4.1), i.e.,

$$\sigma^{-\frac{1}{4}} A_0 \mp \sigma^{\frac{1}{4}} B_0 = \Phi_-^{I,II} + O\left(\frac{\varphi^2}{4\sigma}\right).$$

The relationship of Eq. (4.3) to the second caustic is readily disclosed by invoking relation (1.11). Equation (4.3) may thus be written in the form

$$U_m(x,z) \approx \sqrt[4]{\sigma} \frac{\Gamma(m+1)}{2\Gamma\left(\frac{m}{2}+1\right)} e^{i\frac{\pi}{4}} \left\{ \left(\sigma^{-\frac{1}{4}} A_0 - \sigma^{\frac{1}{4}} B_0\right) e^{i\omega\tau_+^{(2)} - i\frac{\pi}{2}} + \left(\sigma^{-\frac{1}{4}} A_0 + \sigma^{\frac{1}{4}} B_0\right) e^{i\omega\tau_+^{(1)}} \right\}, \tag{4.4}$$

where the expression in the braces again represents the sum of the same two waves. However, as the notation implies, this is the sum of a wave traversing the other caustic $\varphi - 2\sqrt{\sigma} = 0$ and a wave arriving at that same caustic.

Near the caustic $\varphi \pm 2\sqrt{\sigma} = 0$ the first term is the principal term in Eq. (3.22). The parabolic-cylindrical function $\mathcal{D}_m(\omega^{\frac{1}{2}}\varphi)$ can now be approximated by an Airy function, thus concurring with the well-established equations near a single caustic [4, 5, 6].

We conclude with an analysis of how the function $\chi_0^{(-)}(\alpha_-)$ can be specified on some segment $\Delta l < L(\alpha_-)$ of the caustic $\Psi_m^{(-)}=0$, where $L(\alpha_-)$ is the variable involved in the functional equation (2.13).

Let the geometric-optical approximation be valid in some domain Ω interior to the waveguide. Also, let the ray expansion be known for a wave arriving at the caustic $\Psi_m^{(-)}=0$ ([6], Chap. 1):

$$U \sim \sum_{s=0}^{\infty} \frac{U^{(s)}(\mathcal{M})}{(-i\omega)^{s+\gamma}} e^{i\omega\tau_-^{(1)}(\mathcal{M})}, \tag{4.5}$$

where $U^{(s)}(\mathcal{M})$ are known functions of the variable point $\mathcal{M}$ and, in particular,

$$U^{(0)} = \Psi_0(\alpha_-')\sqrt{\frac{c}{J_-'}} .$$

Comparing Eq. (4.3), which is also applicable in Ω, with Eq. (4.5), we have

$$\chi_0^{(-)}(\alpha_-) = \frac{2\Gamma\left(\frac{m}{2}+1\right)}{\sqrt[4]{6}\,\Gamma(m+1)}(-1)^m e^{i\frac{\pi}{2}\left(\gamma-\frac{1}{2}\right)}\Psi_0(\alpha_-).$$

An implication of Eq. (2.13) is the fact that the caustic segment on which the distribution of the function $\chi_0^{(-)}$ can be specified (in general arbitrarily) must not be greater than $L(\alpha_-)$, i.e., the transit time of a perturbation along the caustic $\Psi_m^{(-)}=0$ between two adjacent tangency points of the ray with the caustic.

LITERATURE CITED

1. Yanson, Z. A. High-frequency asymptotic behavior of the wave equation solutions concentrated in a vertically inhomogeneous waveguide, Izv. Akad. Nauk SSSR, Ser. Fizika Zemli (in press).
2. Buldyrev, V. S., The asymptotic behavior of the solutions of the wave equation concentrated near the axis of a two-dimensional waveguide in an inhomogeneous medium, Topics in Mathematical Physics, Vol. 3, Consultants Bureau, New York (1969), pp. 1-23.
3. Babich, V. M., and Alekseev, A. S., On the ray method of calculating wave fronts, Izv. Akad. Nauk SSSR, Ser. Geofiz., No. 1 (1958).
4. Kravtsov, Yu. A., A modification of the method of geometric optics, Izv. VUZov, Radiofizika, Vol. 7, No. 4 (1964).
5. Ludwig, D., Uniform asymptotic expansions at a caustic, Commun. Pure Appl. Math., Vol. 19, No. 3 (1966).
6. Babich, V. M., and Buldyrev, V. S., Asymptotic Methods in Short-Wave Diffraction Problems, Izd. Nauka, Moscow (in press).
7. Keller, J. B., and Rubinaw, S., Asymptotic solution of eigenvalue problems, Ann. Phys., Vol. 9, No. 1 (1960).
8. Kravtsov, Yu. A., Modification of the method of geometric optics for a wave penetrating a caustic, Izv. VUZov, Vol. 8, No. 4 (1965).
9. Bateman, H., and Erdélyi, A., Higher Transcendental Functions, McGraw-Hill, New York (1953).